XoveTIC 2018

XoveTIC 2018

A Coruña, Spain

27–28 September 2018

Issue Editors

Ignacio Fraga
Alberto Alvarellos
Maria Montero
Javier Pereira
Manuel G. Penedo

MDPI • Basel • Beijing • Wuhan • Barcelona • Belgrade

MDPI

Issue Editors

Ignacio Fraga
CITIC - Research Center of Information and
Communication Technologies,
M2NICA Group, Faculty of Informatics,
University of A Coruña, Spain

Alberto Alvarellos
CITIC - Research Center of Information and
Communication Technologies,
RNASA-IMEDIR Group, Faculty of Informatics,
University of A Coruña, Spain

Javier Pereira
CITIC - Research Center of Information and
Communication Technologies,
Faculty of Heath Sciences,
University of A Coruña, Spain

Maria Montero
CITIC - Research Center of Information and
Communication Technologies,
University of A Coruña, Spain

Manuel G. Penedo
CITIC - Research Center of Information and
Communication Technologies, VARPA Group, Faculty of Informatics,
University of A Coruña, Spain

Editorial Office
MDPI, St. Alban-Anlage 66, Basel, Switzerland

This is a reprint of articles from the Issue published online in the open access journal *Proceedings* (ISSN 2504-3900) from 2018 (available at: https://www.mdpi.com/2504-3900/2/18).

For citation purposes, cite each article independently as indicated on the article page online and as indicated below:

LastName, A.A.; LastName, B.B.; LastName, C.C. Article Title. *Journal Name* **Year**, *Article Number*, Page Range.

ISBN 978-3-03897-300-3 (Pbk)
ISBN 978-3-03897-301-0 (PDF)

Cover image courtesy of CITIC - Research Center of Information and Communication Technologies, University of A Coruña, Spain.

Contents

Acknowledgments

Financial support from Consellería de Educación, Universidad y Formación Profesional of the Xunta de Galicia (Convenio I+D+i and Centro Singular de investigación de Galicia accreditation 2016–2019) and the European Union (European Regional Development Fund - ERDF) is gratefully acknowledged.

proceedings

MDPI

Extended Abstract

Raspberry Pimu: Raspberry Pi Based Inertial Sensor Data Processing System [†]

Alberto Alvarellos [1,*], Adrián Vázquez [2] and Juan Rabuñal [2]

[1] Computer Science Department, Research Center on Information and Communication Technologies, University of A Coruña, 15071 A Coruña, Spain
[2] Computer Science Department, Center of Technological Innovations in Construction and Civil Engineering, University of A Coruña, 15071 A Coruña, Spain; adrian.vazqz@gmail.com (A.V.); juanra@udc.es (J.R.)
* Correspondence: alberto.alvarellos@udc.es; Tel.: +34-981-167-000 (ext. 5517)
† Presented at the XoveTIC Congress, A Coruña, Spain, 27–28 September 2018.

Published: 18 September 2018

Abstract: This paper explains the architectural design and development of an application for the reception, visualization and storage of inertial sensor data provided by an inertial measurement system (IMU). The application is built to run in a Raspberry Pi equipped with a small size screen that allows the visualization of the data and the control of data recording. The IMU is connected to a Raspberry Pi through a serial port (USB-TTY).

Keywords: IMU; inertial sensors; Raspberry Pi; Java

1. Introduction

Spain is the European Union country with the longest coastline, with a length of 8000 km. Its geographical location positions it as a strategic element in international shipping and a logistics platform in southern Europe. Events that could disrupt the normal operations of a port, and actions aimed to improve or optimize processes, can have a big economic impact. Port infrastructures are subject to different meteorological conditions (waves, wind, currents ...) that can produce such disruptions. The port must minimize the effect that the meteorological conditions have on ship movements, ensuring that they can operate in a safe manner [1,2].

Port operability is usually quantified based on the movements of moored ships, therefore the lower the impact the meteorological conditions have on ship movements during operations inside the port, the greater the performance of the port is. Our group is currently measuring vessel movements using Inertial Measurement Units [3] and computer vision [4]. The IMUs are also used to validate new computer vision algorithms.

2. System Development

In order for an IMU to be suitable to use in a port environment it should be portable, autonomous and precise. With these characteristics in mind, we developed a system, based on Raspberry Pi and coded in Java, to visualize and record IMU data. The Raspberry Pi is equipped with a small size screen that allows the visualization of the data (see Figure 1). The IMU is connected to a Raspberry Pi through a serial port (USB-TTY).

(a)   (b)

Figure 1. Main screens of the application: (**a**) IMU selection and connection; (**b**) Inertial sensor data visualization and data storage parameters.

The system has been designed to be able to receive data in a precise manner, i.e., the sampling frequency of the IMU needs to be accurate and configurable. This precision is required because the system will be used not only to measure object movements, but also to calibrate and correct computer vision techniques (that allow measuring the movement of objects in a non invasive manner). In Figure 2 we can see an example where the IMU is used to test a Computer Vision based tracking system used to measure the movement of a pendulum.

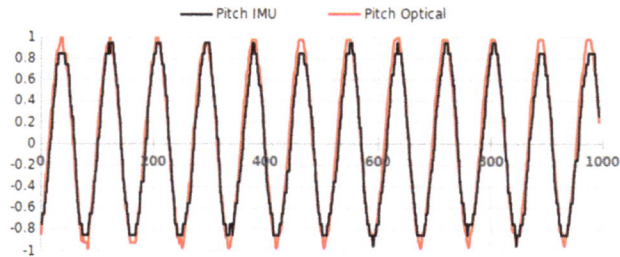

Figure 2. Results of using the IMU to test a Computer Vision tracking algorithm using a pendulum movement (in degrees).

The system is going to be assembled in a water proof case and will be powered by batteries, allowing the system to be autonomous and capable to be used in harsh environments (such as a cargo vessel).

Author Contributions: A.A. designed the system and code architecture (coded the apis and architecture) and wrote the paper; A.V. coded the low level system and tested it; J.R. conceived the system and developed the IMU, based on Arduino + inertial sensors.

Acknowledgments: This research was funded by Xunta de Galicia (Centro singular de investigación de Galicia accreditation 2016–2019) and the European Union (European Regional Development Fund—ERDF).

Conflicts of Interest: The authors declare no conflict of interest. The founding sponsors had no role in the design of the study; in the collection, analyses, or interpretation of data; in the writing of the manuscript, and in the decision to publish the results

References

1. Permanent International Association of Navigation Congresses. *Criteria for Movements of Moored Ships in Harbours: Report of Working Group 24 of the Permanent Technical Committee II*; PIANC: Brussels, Belgium, 1995.
2. Puertos del Estado. *ROM 2.0-11: Recomendaciones Para el Proyecto y Ejecución en Obras de Atraque y Amarre*; Ministerio de Fomento: Madrid, Spain, 2011.
3. Figuero, A.; Rodriguez, A.; Sande, J.; Peña González, E.; Rabuñal, J. Field measurements of angular motions of a vessel at berth: Inertial device application. *Control Eng. Appl. Inform.* **2018**, *20*, in press.
4. Figuero, A.; Rodriguez, A.; Sande, J.; Peña, E.; Rabuñal, J.R. Dynamical Study of a Moored Vessel Using Computer Vision. *J. Mar. Sci. Technol.* **2018**, *26*, 240–250.

proceedings

MDPI

Extended Abstract

Increasing NLP Parsing Efficiency with Chunking †

Mark Dáibhidh Anderson * and David Vilares

FASTPARSE Lab, Departamento de Computación, University of A Coruña, Campus de Elviña, 15071 A Coruña, Spain; david.vilares@udc.es

* Correspondence: m.anderson@udc.es; Tel.: +34-981-167-000

† Presented at the XoveTIC Congress, A Coruña, Spain, 27–28 September 2018.

Published: 19 September 2018

Abstract: We introduce a "Chunk-and-Pass" parsing technique influenced by a psycholinguistic model, where linguistic information is processed not word-by-word but rather in larger chunks of words. We present preliminary results that show that it is feasible to compress linguistic data into chunks without significantly diminishing parsing performance and potentially increasing the speed.

Keywords: Parsing; Syntax; natural language processing; NLP; dependency parsing; Chunking

---

1. Introduction

Syntactic information is required to fully understand linguistic information: utterances are not just a string of words with a meaning solely derived from the semantics of each individual word. The way they are combined also affects meaning. In this context, syntactic analysis can augment many applications in natural language processing (NLP), e.g., state-of-the-art semantic analysis and information retrieval. Dependency parsers are used in these systems as the other main flavour of parsers, constituency parsers, are orders of magnitude slower. Still, state-of-the-art dependency parsers can only process about 100 sentences per second [1]. For large-scale analyses, this is cost-prohibitive.

Dependency parsing represents relations between words with arcs, e.g., the phrase *"I felt"* would have an arc from *"felt"* (the head) to *"I"* (the dependent) and with a *nsubj* label (see Figure 1). Attachment scores are used to evaluate dependency parsers: unlabelled (UAS) measures the number of correct heads and labelled (LAS) measures this and the labelling accuracy.

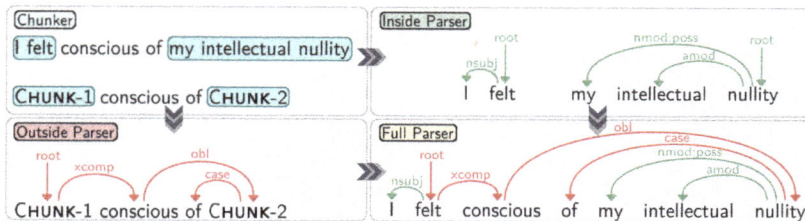

Figure 1. Initially the sentence is processed by the chunker. The contents of the chunks are then sent to the inside parser and the abstract representation of the sentence is sent to the outside parser. The predictions from both are then collated to form the full parse.

Our technique to increase parsing efficiency is inspired by the "Chunk - and - Pass" psycholinguistic model, where an ever increasing abstract hierarchical representation of linguistic input is created in order to process it efficiently and to overcome working-memory restrictions [2]. This entails finding phrases in sentences which can be extracted and processed by a faster but less robust parser, while the more abstract form with more complicated relations is parsed by a slower and more thorough parser.

2. Materials and Methods

The implementation consists of a supervised chunker; an inside parser which analyses the words within the chunks; and an outside parser which analyses the relationships between chunks (see Figure 1). The dataset used was the Universal Dependency English EWT treebank v2.1 [3].

The supervised chunker was implemented using a neural sequence labelling toolkit (NCRF++) [4]. We generated gold labels for the chunker using the BIO tagging scheme, where B is the beginning, I is inside, and O is outside of a chunk. B and I tags were suffixed with the phrase type of the chunk, e.g., B-NP and I-NP for noun-phrase chunks. The labels were generated by using part-of-speech rule sets automatically extracted from the training data. An example rule for a noun phrase could be DET ADJ NOUN. Each set has a threshold on the ratio between invalid (containing unrelated words) and valid chunks when used with an unsupervised rule-based chunker.

The inside parser used the arc eager algorithm in MaltParser [5]. The outside parser used a neural network (NN) implementation of the stack-based arc standard algorithm [6] with universal-dependency-specific features [7]. The inside parser has a speed of ≈16,500 tokens per second (TPS), the chunker of ≈10,200 TPS, and the outside of ≈2000 TPS, so a compression ratio (initial tokens to resulting chunks) of 1.6 can theoretically increase the speed relative to using just the NN parser by 15%.

3. Results

Figure 2a shows the dependency of the supervised chunker's performance on the global ratio threshold of the rule sets used to generate gold-labelled data as described above. Also in Figure 2a the chunker's compression ratio with respect to the rule threshold is shown. Figure 2b shows the parsing performance of the full system, the inside parser, the outside parser, and the corresponding performance of the baseline model (NN stack-based arc standard) for each.

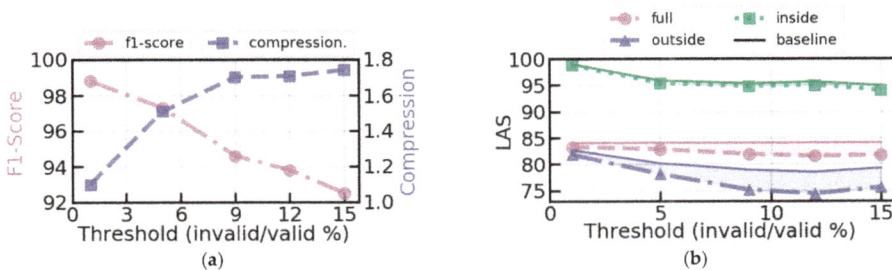

Figure 2. (a) NCRF++ performance and the corresponding compression rate for different rule sets. (b) Inside (green, square), outside (blue, triangle), and full-system (magenta, circle) scores using NCRF++ chunker for different rule sets. Baseline refers to the performances of the baseline model for the corresponding sections that were sent to each sub-parser and are displayed as continuous lines.

4. Discussion

As seen in Figure 2a, it is not useful to use rule sets with ever decreasing performances as the compression return begins to diminish, so there appears to be an upper limit of efficiency improvement. In Figure 2b, it can be observed that the inside chunker does not lose much accuracy. The loss is more pronounced for the outside parser. This is likely due to the decrease in contextual information it has and the more complicated relationships it has to process. Despite this, the best compression to performance rule set (9% threshold) only loses 1.25 UAS and 2.2 LAS points.

Proceedings **2018**, *2*, 1160

We have shown initial results that highlight the efficacy of this approach. Further research will be focused on optimising the implementation and acquiring accurate speed measurements. Beyond this, we will expand the system to process other languages.

Funding: This work has received funding from the European Research Council (ERC), under the European Union's Horizon 2020 research and innovation programme (FASTPARSE, grant agreement No 714150).

References

1. Gómez-Rodríguez, C. Towards fast natural language parsing: FASTPARSE ERC Starting Grant. *Proces. Leng. Nat.* **2017**, *59*, 121–124.
2. Christiansen, M.H.; Chater, N. The Now-or-Never bottleneck: A fundamental constraint on language. *Behav. Brain Sci.* **2016**, *39*, e62, doi:10.1017/S0140525X1500031X.
3. Nivre, J.; Agić, Ž.; Ahrenberg, L.; Antonsen, L.; Aranzabe, M.J.; Asahara, M.; Ateyah, L.; Attia, M.; Atutxa, A.; Augustinus, L.; et al. *Universal Dependencies 2.1*; LINDAT/CLARIN Digital Library at the Institute of Formal and Applied Linguistics (ÚFAL), Faculty of Mathematics and Physics, Charles University: Prague, Czech Republic, 2017.
4. Yang, J.; Zhang, Y. NCRF++: An Open-source Neural Sequence Labeling Toolkit. In Proceedings of the 56th Annual Meeting of the Association for Computational Linguistics, Melbourne, Australia, 15–20 July 2018.
5. Nivre, J.; Hall, J.; Nilsson, J. Maltparser: A data-driven parser-generator for dependency parsing. In Proceedings of LREC, Genoa, Italy, May 2006; pp. 2216–2219.
6. Chen, D.; Manning, C. A fast and accurate dependency parser using neural networks. In Proceedings of the 2014 Conference on Empirical Methods in Natural Language Processing (EMNLP), Doha, Qatar, 2014; pp. 740–750.
7. Straka, M.; Hajic, J.; Straková, J.; Hajic jr, J. Parsing universal dependency treebanks using neural networks and search-based oracle. In Proceedings of the International Workshop on Treebanks and Linguistic Theories (TLT14), Warsaw, Poland, 11–12 December 2015; pp. 208–220.

proceedings

MDPI

Extended Abstract

Automatic Characterization of Epiretinal Membrane in OCT Images with Supervised Training [†]

Sergio Baamonde [1,2,*], Joaquim de Moura [1,2], Jorge Novo [1,2], Noelia Barreira [1,2] and Marcos Ortega [1,2]

[1] VARPA Group, Department of Computer Science, University of A Coruña, 15071 A Coruña, Spain; joaquim.demoura@udc.es (J.d.M.); jnovo@udc.es (J.N.); nbarreira@udc.es (N.B.); mortega@udc.es (M.O.)
[2] CITIC-Research Center of Information and Communication Technologies, University of A Coruña, 15071 A Coruña, Spain
* Correspondence: sergio.baamonde@udc.es; Tel.: +34-656-454-883
† Presented at the XoveTIC Congress, A Coruña, Spain, 27–28 September 2018.

Published: 17 September 2018

Abstract: This work presents an automatic method to characterize the presence or absence of the epiretinal membrane (ERM) in Optical Coherence Tomography (OCT) images. To this end, a predefined set of classifiers is used on multiple local-based feature vectors which represent the inner limiting membrane (ILM), the layer of the retina where the ERM can be present.

Keywords: epiretinal membrane; Retinal Layers; medical imaging; Optical Coherence Tomography

1. Introduction

Optical Coherence Tomography (OCT) is a non-invasive imaging technology which is able to obtain in vivo, cross-sectional and high-resolution images from within the retina. These benefits helped to establish the OCT technique as one of the most widely used techniques for medical imaging. OCT is used in the analysis of pathologies such as glaucoma, Age-related Macular Degeneration (AMD) or Diabetic Macular Edema (DME). Among other eye-related pathologies, OCT imaging can be used to detect the early presence of the epiretinal membrane (ERM) in the surface of the retina, which is crucial to avoid further deterioration, blurring or distortion of the central vision in the affected eye.

This work [1] presents a fully automatic methodology to identify the ERM presence in the OCT images. Other works are focused on the use of manual markers or supervised detections by the specialists, whereas this methodology faces a precise and automatic identification of the region of interest and classification of the points inside this area without the need of any external input.

2. Methodology

The identification of the region of interest (ROI) is done by means of a deformable model which adapts its contour to the ILM layer, area where the ERM can be present.

Once the ILM is identified precisely, we define a feature vector from a local window around each ILM point by applying a feature extraction procedure, as seen on Figure 1a.

Finally, the points of interest inside the ROI are classified using the obtained feature vectors to identify the presence or absence of the epiretinal membrane.

(a)   (b)

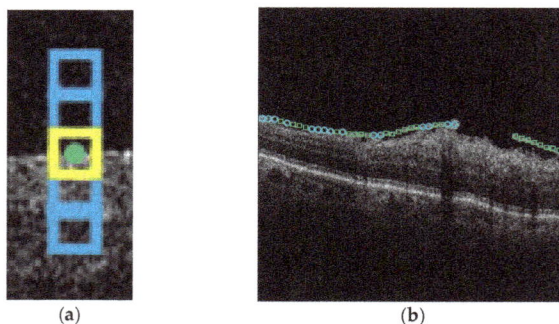

Figure 1. (a) Vertical window around a ROI point. Central region surrounds the analyzed point. (b) Result from the classification process. The circles show the area where the ERM is placed on the ILM, whereas the squares represent the ERM that is separated from the retina.

3. Experimental Results

This methodology was proved by using a dataset of 129 OCT images. 120 samples were equally taken from the complete dataset, highlighting zones with and without ERM presence. Multilayer perceptron, naive Bayes and random forest classifiers were tested to establish the validity of the proposal on top of refining the accuracy and quality of the results. Results (Figure 1b) show the areas with ERM presence, differentiating between areas where the ERM is next to the ILM and areas where the ERM is separated from the retina.

Acknowledgments: This work is supported by the Instituto de Salud Carlos III, Government of Spain and FEDER funds of the European Union through the PI14/02161 and the DTS15/00153 research projects and by the Ministerio de Economía y Competitividad, Government of Spain through the DPI2015-69948-R research project.

Conflicts of Interest: The authors declare no conflict of interest. The founding sponsors had no role in the design of the study; in the collection, analyses, or interpretation of data; in the writing of the manuscript, and in the decision to publish the results.

References

1.   Baamonde, S.; de Moura, J.; Novo, J.; Ortega, M. Automatic Detection of Epiretinal Membrane in OCT Images by Means of Local Luminosity Patterns. *Adv. Comput. Intell.* **2017**, *10305*, 222–235.

proceedings

MDPI

Extended Abstract

A Critical Approach to Information and Communication Technologies †

Alfonso Ballesteros

Philosophy of Law (Private Law Department), Universidad de A Coruña, 15071 A Coruña, Spain;
alfonso.ballesteros@udc.es
† Presented at the XoveTIC Congress, A Coruña, Spain, 27–28 September 2018.

Published: 14 September 2018

Abstract: Many times it has been taken for granted that information and communication technologies (ICT) are intrinsically good for human beings or at least neutral. The first position is assumed by "techno-enthusiasts", the second by those who have a well-meaning opinion of ICT. Here we briefly framed a third possibility leaded by South-Korean philosopher Byung-Chul Han, a position that allows us to think about how ICT is shaping society and human beings as we know it.

Keywords: information and communication technologies; positive society; Transparency Society

1. Introduction

Since their appearance it has been usually taken for granted that information and communication technologies (ICT) are good for human beings or at least neutral. The first position is assumed by "techno-enthusiasts", the second by those who have a well-meaning opinion of ICT. All in all those views are not the only ones and have been challenged recently. Some philosophers are reluctant to consider information and communication technologies positive or even neutral. Here we frame this position leaded by Byung-Chul Han.

2. Discussion

If information and communication technologies are neutral they are not good or bad themselves but the use of them could be considered good or bad. This position excludes any responsibility of those who produce these technologies and it makes the ICT user the responsible alone. This view forgets that human actions are mediated and partially determined by objects, tools and technologies (1). It also forgets that objects like a chair or a smart-phone are far from having the same influence in human beings (2). As objects and tools become more and more complex are designed not by craft makers but by prestigious and clever engineers. If we think twice, a chair made by a carpenter and a smart-phone made by a group of engineers force us to have a very different perspective and a different judgment on our relations with technology.

Leading South-Korean philosopher Byung-Chul Han has underlined the negative effects of digital technology in society. Han defines our society as a "positive society", but this is not positive in the sense of good. By positive he means that we live in an immature society unable to face reality and especially what is hard or painful: illnesses, death, ugliness or even disagreement [1] (pp. 11–23).

Han also states that Information and Communication Technologies lead not to more communication between people but, on the contrary, to incapacity to listen, narcissism, loneliness and depression. Those are some of the features of what he calls "homo digitalis". With respect of narcissism and the loss of the principle of reality Han's source is Sigmund Freud and his well-known distinction between the reality principle and that of pleasure. The digital realm keeps reality distant and puts the individual and its pleasure in the center. Websites, apps, screens in general, are completely adapted to the tastes of each user. Tastes known by the clicks, likes or any information recorded about that person.

Han also analyses the digital realm as a manifestation of developed capitalism where everything is in plain-sight (as a shop window) and everything has a price. As a consequence transparency and homogeneity become the rule. In a market society money can buy everything and even people become products. People are obsessed of their image, fame and online social recognition. Thus every detail of their lives should be in plain-sight to be "consumed", with the obvious consequence of the loss of privacy and a not real interaction.

Against those negative effects of technology Han's proposal is the recovery of "distance", the recognition of the other as a different person out of our control [1] (p. 16). A person we cannot "delete" from reality although we can delete him or her in social media.

We do not need to accept completely Byung-Chul Han's philosophy to learn from it how to understand technology. In my opinion we do not need to get technology out of our lives, but put it in the real service of human beings. We need a "technology with a human face" as Schumacher, following Gandhi, expressed it [2] (p. 126). The recovery of a direct and real relationship with others in a low-tech environment seems critical and this point becomes stronger as we realized that Silicon Valley parents are raising their kids tech-free and are not precisely techno-enthusiasts [3].

Author Contributions: All work was done by Alfonso Ballesteros.

Conflicts of Interest: The author declares no conflict of interests.

References

1. Han, B.-C. *La Sociedad de la Transparencia*; Gabás, R., Ed.; Herder: Barcelona, Spain, 2013.
2. Schumacher, E.F. *Small is Beautiful: Economics as If People Mattered*; Blond & Briggs: London, UK, 1973.
3. Business Insider. Available online: https://www.businessinsider.com/silicon-valley-parents-raising-their-kids-tech-free-red-flag-2018-2 (accessed on 7 September 2018).

proceedings

MDPI

Extended Abstract

Image Transmission: Analog or Digital? [†]

Jose Balsa *, Tomás Domínguez-Bolano, Óscar Fresnedo, José A. García-Naya and Luis Castedo

Enxeñaría de Computadores, Facultade de Informática, Universidade da Coruña, Campus de Elviña s/n, 15071 A Coruña, Spain; tomas.bolano@udc.es (T.D.-B.); oscar.fresnedo@udc.es (Ó.F.); jagarcia@udc.es (J.A.G.-N.); luis@udc.es (L.C.)
* Correspondence: j.balsa@udc.es
† Presented at the XoveTIC Congress, A Coruña, Spain, 27–28 September 2018.

Published: 18 September 2018

Abstract: Evaluation and comparison of analog and digital wireless transmission systems.

Keywords: analog image transmission; JSCC; image quality comparison; analog image transformation

1. Introduction

In this work, we address the design and evaluation of an analog Joint Source Channel Coding (JSCC) system [1,2] for the transmission of grayscale still images. This kind of systems offers some interesting characteristics with respect to digital systems under certain circumstances: high transmission rates over noisy channels and avoiding retransmissions, graceful degradation with a fixed coding scheme (i.e., if the transmitter does not know the quality of the channel) and lower computational complexity.

The second part of this work evaluates the proposed analog system and compares its performance to that of an all-digital system based on JPEG compression. It is very important the selection of the metrics used to evaluate the perceived quality of the received image respect to that of the transmitted one and defining an adequate strategy to make a fair comparison between the analog and digital systems.

2. System Model

The transmission system consists of three parts: the transmitter, where the image is prepared, adapted and encoded; the channel; and the receiver, where the image is recovered by two operations: undoing the operations carried out in the transmitter and trying to correct the errors due to the distortions introduced by the wireless channel.

Orthogonal Frequency Division Multiplexing (OFDM) is selected as the waveform for both the analog and the all-digital systems. The latter also encodes the data frames using Turbo Codes. The wireless channel can be (1) a simulated one considering a channel model defined by the ITU or (2) a real one obtained by means of over-the-air transmissions with the GTEC testbed [3–6].

2.1. Image Pre-Processing

The image is loaded from a grayscale file with values from 0 to 255 (grayscale intensity). The system divides the image in 8 × 8 blocks and they are transformed with a DCT (Discrete Cosine Transform) (Figure 1a) to concentrate most of the image information into a few coefficients (Figure 1b), such that the least relevant coefficients can be disregarded with a small loss in terms of perceived image quality. Each 8 × 8 block is further subdivided into four blocks, from the first to the third will be transformed (see Section 2.2), whereas the fourth is always discarded.

After this, the system rearranges the transformed blocks into a sequential data frame (Figure 1c) which is ready for the next step.

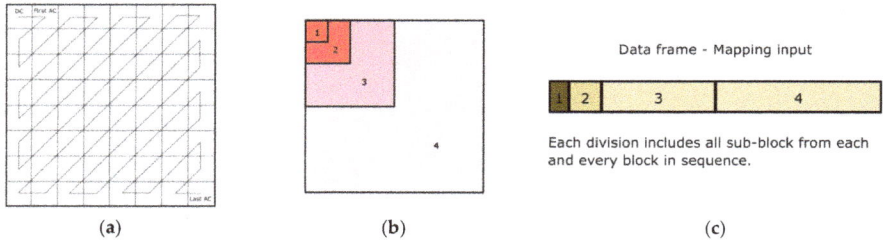

Figure 1. (**a**) The DCT is applied to each 8 × 8 pixel block. (**b**) Each 8 × 8 block is split into 4 sub-blocks in order to apply different analog mapping/transformation schemes. (**c**) Each sub-block is placed in a sequential data frame.

2.2. Analog JSCC Scheme

The system maps the DCT coefficients of the three first blocks (Figure 1c) in channel symbols using analog JSCC techniques [7]. This strategy represents an appealing alternative to the traditional all-digital approaches based on source-channel separation because of its lower complexity and graceful degradation against variations of the channel conditions [2,8]. The system works with two analog JSCC methods: expansion and uncoded. The expansion method applies an analog spherical code, based on the exponentially chirped modulation (Equation (1)), whereas the uncoded method performs only a normalization.

$$ x = \Delta \left[\cos(2\pi s), \sin(2\pi s), \cos(2\pi\alpha s), \sin(2\pi\alpha s), \dots, \cos\left(2\pi\alpha^{\left(\frac{\rho}{2}-1\right)}s\right), \operatorname{sen}\left(2\pi\alpha^{\left(\frac{\rho}{2}-1\right)}s\right) \right] \qquad (1) $$

Equation (1). Analog spherical code used in the encoding operation
s: symbols at the mapping input, α: mapping parameter; Δ normalization factor; ρ: expansion factor.

At the receiver, we use an optimal Minimum Mean Square Error (MMSE) or a Maximum Likelihood (ML) decoder depending on the selected scheme.

2.3. Comparison Methodology between Digital and Analog Systems

The main objective of this work is to compare the transmission speed between the proposed analog system and an all-digital one used as a benchmark. One cannot find in the literature a standard procedure to carry out this very specific comparison. To ensure a fair comparison, we first measure the image quality obtained with the analog system, and next, we design the digital JPEG-based system to compress the source image with that same quality, while ensuring a free-error transmission of the information.

The digital transmission is software-simulated over the estimated channel in the analog system in order to compare the two systems in the same situation. In addition, knowing the channel before transmitting allows for calculating the optimal CQI (Channel Quality Indicator) obtaining the minimal data frame avoiding errors.

The selected quality metric is the SSIM (Structural Similarity Index) [9]. This metric returns a real value 1 (identical images) or less than 1 (different images).

The time needed to transmit the data frames containing the image, encoded with the digital or the analog scheme, is the considered performance metric, which measures the transmission speed.

Acknowledgments: This work has been funded by the Xunta de Galicia (ED431C 2016-045, ED341D R2016/012, and ED431G/01), the Agencia Estatal de Investigacion of Spain (TEC2015-69648-REDC, and TEC2016-75067-C4-1-R), ERDF funds of the EU (AEI/FEDER, UE), and the predoctoral Grant BES-2014-069772.

Conflicts of Interest: The authors declare no conflict of interest. The founding sponsors had no role in the design of the study; in the collection, analyses, or interpretation of data; in the writing of the manuscript, and in the decision to publish the results.

Proceedings **2018**, *2*, 1163

References

1. Ramstad, T.A. Shannon mappings for robust communication. *Telektronikk* **2002**, *98*, 114–128.
2. Hekland, F.; Floor, P.A.; Ramstad, T.A. Shannon-Kotel'nikov mappings in joint source-channel coding. *IEEE Trans. Commun.* **2009**, *57*, 94–105.
3. Domínguez-Bolaño, T.; Rodríguez-Piñeiro, J.; García-Naya, J.A.; Yin, X.; Castedo, L. Measurement-Based Characterization of Train-to-Infrastructure 2.6 GHz Propagation Channel in a Modern Subway Station. *IEEE Access* **2018**.
4. Domínguez-Bolaño, T.; Rodríguez-Piñeiro, J.; García-Naya, J.A.; Castedo, L. Experimental Characterization of LTE Wireless Links in High-Speed Trains. *Wirel. Commun. Mob. Comput.* **2017**, 5079130, doi:10.1155/2017/5079130.
5. Zhang, L.; Rodríguez-Piñeiro, J.; Fernández, J.R.O.; García-Naya, J.A.; Matolak, D.W.; Briso, C.; Castedo, L. Propagation Modeling for Outdoor-to-Indoor and Indoor-to-Outdoor Wireless Links in High-Speed Train. *Measurement* **2017**, *110*, 43–52.
6. Suárez-Casal, P.; Rodríguez-Piñeiro, J.; García-Naya, J.A.; Castedo, L. Experimental Evaluation of the WiMAX Downlink Physical Layer in High-Mobility Scenarios. *EURASIP J. Wirel. Commun. Netw.* **2015**, *109*, doi:10.1186/s13638-015-0339-9.
7. Fresnedo, O.; Vazquez-Araujo, F.J.; Castedo, L.; Garcia-Frias, J. Low-complexity near-optimal decoding for analog joint source channel coding using space-filling curves. *IEEE Commun. Lett.* **2013**, *17*, 745–748.
8. Vaishampayan, V.A.; Costa, S.I.R. Curves on a sphere, shift-map dynamics, and error control for continuous alphabet sources. *IEEE Trans. Inf. Theory* **2003**, *49*, 1658–1672.
9. Wang, Z.; Bovik, A.C. Image Quality Assessment: From Error Visibility to Structural Similarity. *IEEE Trans. Image Process.* **2004**, *13*, doi:10.1109/TIP.2003.819861.

proceedings

MDPI

Extended Abstract

Computationally Efficient Bootstrap Expressions for Bandwidth Selection in Nonparametric Curve Estimation [†]

Inés Barbeito * and Ricardo Cao

Research group MODES, Department of Mathematics, CITIC, Universidade da Coruña, 15071 A Coruña, Spain; rcao@udc.es
* Correspondence: ines.barbeito@udc.es; Tel.: +34-881-011-301
† Presented at the XoveTIC Congress, A Coruña, Spain, 27–28 September 2018.

Published: 17 September 2018

check for updates

Abstract: Bootstrap methods are used for bandwidth selection in: (1) nonparametric kernel density estimation with dependent data (smoothed stationary bootstrap and smoothed moving blocks bootstrap), and (2) nonparametric kernel hazard rate estimation (smoothed bootstrap). In these contexts, four new bandwidth parameter selectors are proposed based on closed bootstrap expressions of the MISE of the kernel density estimator (case 1) and two approximations of the kernel hazard rate estimation (case 2). These expressions turn out to be very useful since Monte Carlo approximation is no longer needed. Finally, these smoothing parameter selectors are empirically compared with the already existing ones via a simulation study.

Keywords: hazard rate; Kernel Method; Mean integrated squared error; moving blocks bootstrap; Smooth Bootstrap; smoothing parameter; stationary bootstrap; Stationary Processes

1. Introduction

This work deals with the well known problem of data-driven choice of smoothing parameters in nonparametric density and hazard rate estimation (see [1–4]). Our aim is also to propose new bootstrap procedures for nonparametric density estimation considering dependent data. On the other hand, hazard rate estimation is considered and two bootstrap bandwidth selectors based on some approximation of the kernel hazard rate estimator are proposed.

2. Nonparametric Density Estimation

Let us consider a random sample, $(X_1, \ldots, X_n)$, coming from a population with density f and the kernel density estimator (see [5,6]), which strongly depends on a bandwidth selector, h. In fact, its choice is really important since it regulates the degree of smoothing applied to the data.

In this context, the smoothed stationary bootstrap (SSB) resampling plan has been proposed (see the Appendix for a detailed description of the algorithm and [7]), as well as a bandwidth selector, namely h^*_{SSB}. It is the result of minimizing the SSB version of the MISE. A closed expression for the bootstrap MISE is also obtained by [7]. On the other hand, smoothed moving blocks bootstrap (SMBB) has been proposed (see the Appendix for a complete description of the method), as well as a bandwidth selector, h^*_{SMBB}, which is the minimizer in h of the closed expression for the $MISE^*_{SMBB}$ (see [8] for a deeper insight on the topic). It is worth mentioning that the exact expressions for the $MISE^*_{SSB}(h)$ and $MISE^*_{SMBB}(h)$ are really useful since Monte Carlo approximation is no longer necessary.

3. Nonparametric Hazard Rate Estimation

Let us consider $(X_1, X_2, \ldots, X_n)$, a simple random sample coming from a population with continuous density f and cumulative distribution function F. Consider, additionally, the nonparametric hazard rate estimator (see [3,4]), the kernel density estimator $\hat{f}_h$ and the kernel distribution estimator $\hat{F}_h$. In order to establish a bootstrap bandwidth selector for the hazard rate estimator, two approximations of the hazard rate estimator are considered. The two hazard rate approximated versions are given by:

$$\tilde{r}_{h,1}(x) = \frac{\hat{f}_h(x)}{1 - F(x)}.$$

$$\tilde{r}_{h,2}(x) = \frac{1}{1 - F(x)} \hat{f}_h(x) + \frac{f(x)}{(1 - F(x))^2} \hat{F}_h(x) - \frac{f(x)}{(1 - F(x))^2} + r(x).$$

Closed-form expressions of the MISE of $\tilde{r}_{h,1}$ and $\tilde{r}_{h,2}$, as well as their bootstrap versions can be found in [9]. Moreover, two bootstrap bandwidth selectors, namely h_{BOOT1} and h_{BOOT2}, are defined as the minimizers of $MISE^*_{\tilde{r}_{h,1},w}(h)$ and $MISE^*_{\tilde{r}_{h,2},w}(h)$, respectively (see [9] for a deeper insight on the approach). It is worth mentioning that Monte Carlo approximation is not required.

4. Simulation Results

A simulation study is now carried out in order to check the good empirical behaviour of the new smoothing parameter selectors in both contexts. These are the models considered:

1. **Density estimation:** An AR(1) model given by $X_t = -0.6X_{t-1} + 0.8a_t$, where $a_t \overset{d}{=} N(0, 1)$.
2. **Hazard rate estimation:** A Gumbel model such that $f(x) = e^{-x} e^{-e^{-x}}, \forall x \geq 0$.

5. Discussion

Figure 1 shows that h^*_{SSB} and h^*_{SMBB} display a similar performance, actually the best one. According to Table 1, h_{BOOT1} and h_{BOOT2} display the overall best performance.

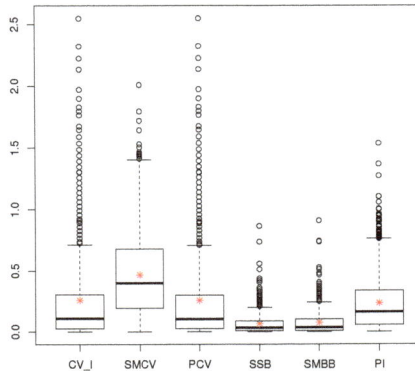

Figure 1. Boxplot of $\log\left(MISE(\hat{h})/MISE(h_{MISE})\right)$, $n = 100$, where $\hat{h} = h_{CV_i}$ (first box), h_{SMCV} (second box), h_{PCV} (third box), h^*_{SSB} (fourth box), h^*_{SMBB} (fifth box) and h_{PI} (sixth box).

Table 1. Mean and median of $ISE(\hat{h})$, $n = 100$, where $\hat{h} = h_{CV}$ (third column), h_{DO} (fourth column), h_{BOOT1} (fifth column), h_{BOOT2} (sixth column) and h_{GCM}^* (seventh column).

		CV	DO	BOOT1	BOOT2	GCM
Gumbel model	Mean	0.1656	0.01651	0.02914	0.02882	0.03595
	Median	0.15527	0.01037	0.012844	0.01282	0.01739

Funding: The authors acknowledge partial support by MINECO grants MTM2014-52876-R and MTM2017-82724-R (EU ERDF support included). Additionally, financial support from the Xunta de Galicia (Centro Singular de Investigación de Galicia accreditation ED431G/01 2016-2019 and Grupos de Referencia Competitiva ED431C2016-015) and the European Union (European Regional Development Fund - ERDF), is gratefully acknowledged. The first author aknowledges financial support from the Xunta de Galicia and the European Union (European Social Fund - ESF), the reference of which is ED481A-2017/215. Additionally, the work of the first author has been partially carried out during a visit at the University of California, San Diego, financed by INDITEX, with reference INDITEX-UDC 2017.

Conflicts of Interest: The authors declare no conflict of interest. The founding sponsors had no role in the design of the study; in the collection, analyses, or interpretation of data; in the writing of the manuscript, and in the decision to publish the results.

Abbreviations

The following abbreviations are used in this manuscript:

MISE   Mean integrated squared error
ISE   Integrated squared error
SSB   Smoothed stationary bootstrap
SMBB   Smoothed moving blocks bootstrap
iid   Independent and identically distributed
h_{DO} DO-validation bandwidth selector for hazard rate estimation (see [10])
h_{GCM}^* González-Manteiga, Cao, Marron bandwidth selector for hazard rate estimation (see [11])
h_{PI} Plug-in bandwidth selector for bandwidth selection with dependent data (see [12])
h_{CV_l} Leave-$(2l + 1)$-out cross-validation for density estimation (see [13])
h_{SMCV} Modified cross validation for density estimation with dependent data (see [8])
h_{PCV} Penalized cross validation for density estimation with dependent data (see [8])
h_{CV} Cross validation bandwidth selector for hazard rate estimation (see [14])
h_{MISE} Bandwidth selector which minimizes the theoretical MISE(h)

Appendix A

Smoothed stationary bootstrap

1. Draw $X_1^{*(SB)}$ from F_n, the empirical distribution function of the sample.

2. Define $X_1^* = X_1^{*(SB)} + gU_1^*$, where U_1^* has been drawn with density K and independently from $X_1^{*(SB)}$.

3. Assume we have already drawn $X_1^*, \ldots, X_i^*$ (and, consequently, $X_1^{*(SB)}, \ldots, X_i^{*(SB)}$) and consider the index j, for which $X_i^{*(SB)} = X_j$. We define a binary auxiliary random variable I_{i+1}^*, such that $P^*\left(I_{i+1}^* = 1\right) = 1 - p$ and $P^*\left(I_{i+1}^* = 0\right) = p$. We assign $X_{i+1}^{*(SB)} = X_{(j \bmod n)+1}$ whenever $I_{i+1}^* = 1$ and we use the empirical distribution function for $X_{i+1}^{*(SB)}|_{I_{i+1}^*=0}$, where mod stands for the modulus operator.

4. Once drawn $X_{i+1}^{*(SB)}$, we define $X_{i+1}^* = X_{i+1}^{*(SB)} + gU_{i+1}^*$, where, again, U_{i+1}^* has been drawn from the density K and independently from $X_{i+1}^{*(SB)}$.

Smoothed moving blocks bootstrap

1. Fix the block length, $b \in \mathbb{N}$, and define $k = \min_{\ell \in \mathbb{N}} \ell \geq \frac{n}{b}$

2. Define:

$$B_{i,b} = (X_i, X_{i+1}, \ldots, X_{i+b-1})$$

3. Draw $\xi_1, \xi_2, \ldots, \xi_k$ with uniform discrete distribution on $\{B_1, B_2, \ldots, B_q\}$, with $q = n - b + 1$

4. Define $X_1^{*(MBB)}, \ldots, X_n^{*(MBB)}$ as the first n components of

$$(\xi_{1,1}, \xi_{1,2}, \ldots, \xi_{1,b}, \xi_{2,1}, \xi_{2,2} \ldots, \xi_{2,b}, \ldots, \xi_{k,1}, \xi_{k,2}, \ldots, \xi_{k,b})$$

5. Define $X_i^* = X_i^{*(MBB)} + gU_i^*$, where U_i^* has been drawn with density K and independently from $X_i^{*(MBB)}$, for all $i = 1, 2, \ldots, n$

References

1. Silverman, B.W. *Density Estimation for Statistics and Data Analysis*; Chapman & Hall: London, UK, 1986.
2. Devroye, L. *A Course in Density Estimation*; Birkhauser: Boston, MA, USA, 1987.
3. Watson, G.S.; Leadbetter, M.R. Hazard analysis I. *Biometrika* **1964a**, *51*, 175–184.
4. Watson, G.S.; Leadbetter, M.R. Hazard analysis II. *Sankhyā Ser. A* **1964b**, *26*, 101–116.
5. Parzen, E. Estimation of a probability density-function and mode. *Ann. Stat.* **1962**, *33*, 1065–1076.
6. Rosenblatt, M. Estimation of a probability density-function and mode. *Ann. Stat.* **1956**, *27*, 832–837.
7. Barbeito, I.; Cao, R. Smoothed stationary bootstrap bandwidth selection for density estimation with dependent data. *Comput. Stat. Data Anal.* **2016**, *104*, 130–147.
8. Barbeito, I.; Cao, R. A review and some new proposals for bandwidth selection in nonparametric density estimation for dependent data. In *From Statistics to Mathematical Finance: Festschrift in Honour of Winfried Stute*; Ferger, D., González Manteiga, W., Schmidt, T., Wang, J.L., Eds.; Springer International Publishing: Cham, Switzerland, 2017; pp. 173–208, ISBN 978-3-319-50986-0.
9. Barbeito, I.; Cao, R. Smoothed bootstrap bandwidth selection for nonparametric hazard rate estimation. *Preprint* **2018**.
10. Gámiz, M.L.; Mammen, E.; Martínez-Miranda, M.D.; Nielsen, J.P. Double one-sided cross-validation of local linear hazards. *J. R. Stat. Soc. Ser. B Stat.* **2016**, *78*, 775–779.
11. González-Manteiga, W.; Cao, R.; Marron, J.S. Bootstrap Selection of the Smoothing Parameter in Nonparametric Hazard Rate Estimation. *J. Am. Stat. Assoc.* **1996**, *91*, 1130–1140.
12. Hall, P.; Lahiri, S.N.; Truong, Y.K. On bandwidth choice for density estimation with dependent data. *Ann. Stat.* **1995**, *23*, 2241–2263.
13. Hart, J.D.; Vieu, P. Data-driven bandwidth choice for density estimation based on dependent data. *Ann. Stat.* **1990**, *18*, 873–890.
14. Patil, P.N. On the Least Squares Cross-Validation Bandwidth in Hazard Rate Estimation. *Ann. Stat.* **1993**, *21*, 1792–1810.

proceedings

MDPI

Extended Abstract

An R Package Implementation for Statistical Modeling of Emergence Curves in Weed Science [†]

Daniel Barreiro-Ures [*,‡], **Ricardo Cao** [‡] **and Mario Francisco-Fernández** [‡]

Department of Mathematics, Faculty of Computer Science, University of A Coruña, 15008 A Coruña, Spain; ricardo.cao@udc.es (R.C.); mario.francisco@udc.es (M.F.-F.)

*   Correspondence: daniel.barreiro.ures@udc.es
†   Presented at the XoveTIC Congress, A Coruña, Spain, 27–28 September 2018.
‡   These authors contributed equally to this work.

Published: 18 September 2018

Abstract: Over the last few years, the research group MODES has carried out a research line (in collaboration with researchers from the Sustainable Agriculture Institute of the CSIC in Córdoba) on statistical modeling in weed science. One of the aspects dealt with in this line is that of the estimation of the so-called emergence curves from data obtained from field studies. In this context, new indices have been developed for hydrothermal times, new nonparametric methods have been proposed, which have been compared with other existing parametric methods and applied to relevant pests. In this context, the objective pursued was the development of an R package that can be useful for the statistical analysis of weed science data and, in particular, for the estimation of emergence curves. Currently, the package is available in the CRAN and it is intended to become a standard of use among the research community in weed science.

Keywords: binned data; nonparametric; kernel density; kernel distribution; bandwidth selection

1. Scenario

Let $X_1, \ldots, X_n$ be a random sample of our random variable of interest, X, with density function f and distribution function F. Let us assume that a set of k intervals, $[y_{j-1}, y_j)$, $j = 1, \ldots, k$, whose midpoints will be denoted by $t_j = \frac{1}{2}(y_{j-1} + y_j)$, $j = 1, \ldots, k$. Let us also assume that we do not know the values of each X_i but only the interval to which each of them belongs. Therefore, in this scenario we cannot directly observe the sample $X_1, \ldots, X_n$ but only the proportion of observations that fall into each of the intervals, that is, our random sample turns to be $(w_1, \ldots, w_k)$, where $w_j = F_n(y_j^-) - F_n(y_{j-1}^-)$, $j = 1, \ldots, k$, and F_n denotes the empirical distribution function of $X_1, \ldots, X_n$. In this context, the classical kernel density and distribution estimators cannot be computed and must be adapted to this interval-grouping scenario. Kernel density and ditribution estimators for interval-grouped data were proposed in [1,2], respectively.

2. Bandwidth Selection for Interval-Grouped Data

The bandwidth selectors proposed in [3] were analyzed through several simulation studies and some modifications were proposed. For instance, in the case of kernel density estimation, a new method for the selection of a pilot bandwidth for the bootstrap selector was proposed and shown to outperform the previously used one in most scenarios. In the case of kernel distribution estimation, a bootstrap bandwidth selector was developed along with a method to select a pilot bandwidth similar to the one proposed for the density case. For both kernel density and distribution estimation, the bootstrap bandwidth selectors were shown to outperform the plug-in selectors in most cases.

Proceedings **2018**, *2*, 1165

3. Application to the Estimation of Seedling Emergence Curves

Weed scientists are usually interested in the prediction of seedling emergence using environmental variables such as the cumulative hydrothermal time (CHTT). However, due to several factors such as budget constraints, the value of the CHTT cannot be measured continuously and so we end up facing an interval-grouped scenario. Weed scientists have traditionally modeled the relationship between seedling emergence and CHTT through parametric regression. To overcome the limitations of this approach, we decided to face the task of seedling emergence estimation from a nonparametric distribution estimation viewpoint. Namely, we want to estimate the distribution (or density) of the random variable CHTT. Furthermore, due to the nature of the measuring process and the fact that the seedlings under study are at different soil depths, we also face the problem of selecting the depth at which to measure the CHTT. Since the depth at which the CHTT is measured will affect the shape of the distribution of our random variable, our objective is to find a depth such that it maximizes the flatness of the distribution of the CHTT. For this task, emergence indices were proposed in [4].

4. Implementation

The methods were coded in C++ to minimize the execution time and integrated into an R package, binnednp [5], through the Rcpp API (see [6]). The package is composed of four functions focused on different tasks: kernel density estimation (*bw.dens.binned*), plug-in bandwidth selection for kernel distribution estimation (*bw.dist.binned*), bootstrap bandwidth selection for kernel distribution estimation (*bw.dist.binned.boot*) and nonparametric estimation of emergence indices (*emergence.indices*).

Author Contributions: Conceptualization, D.B., R.C. and M.F.; Methodology, D.B., R.C. and M.F.; Software, D.B., R.C. and M.F.; Validation, D.B., R.C. and M.F.; Formal Analysis, D.B., R.C. and M.F.; Investigation, D.B., R.C. and M.F.; Resources, D.B., R.C. and M.F.; Data Curation, D.B., R.C. and M.F.; Writing—Original Draft Preparation, D.B., R.C. and M.F.; Writing—Review & Editing, D.B., R.C. and M.F.; Visualization, D.B., R.C. and M.F.; Supervision, D.B., R.C. and M.F.; Project Administration, D.B., R.C. and M.F.; Funding Acquisition, D.B., R.C. and M.F.

Funding: This research received no external funding.

Acknowledgments: This research has been supported by MINECO grant MTM-2014-52876-R and by the Xunta de Galicia (Grupos de Referencia Competitiva ED431C-2016-015 and Centro Singular de Investigación de Galicia ED431G/01), all of them through the ERDF.

References

1. Reyes, M.; Francisco-Fernández, M.; Cao, R. Nonparametric kernel density estimation for general grouped data. *J. Nonparametr. Stat.* **2016**, *2*, 235–249.
2. Cao, R.; Francisco-Fernández, M.; Anand, A.; Bastida, F.; González-Andújar, J.L. Modeling *bromus diandrus* seedling emergence using nonparametric estimation. *J. Agric. Biol. Environ. Stat.* **2013**, *18*, 64–86.
3. Reyes, M.A. Statistical Methods for Studying Emergence Curves in Weed Science. Ph.D. Thesis, Universidade da Coruña, A Coruña, Spain, 2015.
4. Cao, R.; Francisco-Fernández, M.; Anand, A.; Bastida, F.; González-Andújar, J.L. Computing statistical indices for hydrothermal times using weed emergence data. *J. Agric. Biol. Environ. Stat.* **2011**, *149*, 701–712.
5. Barreiro-Ures, D.; Fraguela, B.; Doallo, R.; Cao, R.; Francisco-Fernández, M.; Reyes, M. binnednp: Nonparametric Estimation for Interval-Grouped Data. *CRAN* **2018**. Available online: https://cran.r-project.org/package=binnednp,Rpackageversion0.1.0 (accessed on 9 September 2018).
6. Eddelbuettel, D.; Francois, R. Rcpp: Seamless R and C++ integration. *J. Stat. Softw.* **2011**, *40*, 1–18.

proceedings

MDPI

Extended Abstract

Bandwidth Selection in Nonparametric Regression with Large Sample Size [†]

Daniel Barreiro-Ures [‡], Ricardo Cao [‡] and Mario Francisco-Fernández [‡]

Department of Mathematics, Faculty of Computer Science, University of A Coruña, A Coruña 15008, Spain; ricardo.cao@udc.es (R.C.); mario.francisco@udc.es (M.F.-F.)

* Correspondence: daniel.barreiro.ures@udc.es

† Presented at the XoveTIC Congress, A Coruña, Spain, 27–28 September 2018.

‡ These authors contributed equally to this work.

Published: 17 September 2018

Abstract: In the context of nonparametric regression estimation, the behaviour of kernel methods such as the Nadaraya-Watson or local linear estimators is heavily influenced by the value of the bandwidth parameter, which determines the trade-off between bias and variance. This clearly implies that the selection of an optimal bandwidth, in the sense of minimizing some risk function (MSE, MISE, etc.), is a crucial issue. However, the task of estimating an optimal bandwidth using the whole sample can be very expensive in terms of computing time in the context of Big Data, due to the computational complexity of some of the most used algorithms for bandwidth selection (leave-one-out cross validation, for example, has $\mathcal{O}(n^2)$ complexity). To overcome this problem, we propose two methods that estimate the optimal bandwidth for several subsamples of our large dataset and then extrapolate the result to the original sample size making use of the asymptotic expression of the MISE bandwidth. Preliminary simulation studies show that the proposed methods lead to a drastic reduction in computing time, while the statistical precision is only slightly decreased.

Keywords: nonparametric; regression; bandwidth; Big Data; cross-validation; subsampling

---

1. Scenario

Let us consider a sample of size n, $\{(x_i, y_i)\}_{i=1,...,n}$, drawn from a nonparametric regression model $y_i = m(x_i) + \varepsilon_i$. We assume random design, $\mathbb{E}[\varepsilon \mid x] = 0$ and $\mathbb{E}[\varepsilon^2 \mid x] = \sigma^2(x) < \infty$. In this context, we deal with the Nadaraya-Watson estimator [1] for the regression function, m, which is characterized by the kernel function K and the bandwidth or smoothing parameter $h > 0$. Under suitable conditions, the asymptotically optimal (in the sense of minimum AMISE) bandwidth satisfies

$$h_{AMISE,n} = c_0 n^{-\frac{1}{5}}. \tag{1}$$

Since we are assuming that the sample size, n, is very large, the task of computing a bandwidth selector using the whole sample would be too computationally expensive. For example, the leave-one-out cross-validation (LOO CV) bandwidth selector has complexity $\mathcal{O}(n^2)$.

2. Bandwidth Selection

The idea behind our proposal is to find the LOO CV bandwidth for several subsamples and then extrapolate the result to the original sample size using the asymptotic expression of the MISE bandwidth (1).

2.1. One Subsample Size (OSS)

The idea behind this method is to draw several subsamples of size r, much smaller than n, then compute the LOO CV selector and finally use Equation (1) to extrapolate the CV bandwidth for the original sample size (this idea was already proposed in [2] in the context of kernel density estimation to reduce the variance of the CV bandwidth selector).

1. Obtain s subsamples of size $r \ll n$ subsampling without replacement from our original dataset.
2. For each subsample, find the LOO CV bandwidth.
3. Let $\hat{h}_r$ denote the average of these bandwidths.
4. We estimate the unknown constant c_0 by $\hat{c}_0 = \hat{h}_r r^{\frac{1}{5}}$.
5. Therefore, our estimate of the AMISE bandwidth would be $\hat{h}_{AMISE,n} = \hat{c}_0 n^{-\frac{1}{5}} = \hat{h}_r \left(\frac{r}{n}\right)^{\frac{1}{5}}$.

2.2. Several Subsample Sizes (SSS)

We now propose a method that considers several subsamples of different sizes.

1. Consider a grid of subsample sizes, $r_1, \ldots, r_s$, with $r_j \ll n$.
2. For each r_j, compute the LOO CV bandwidth, $\hat{h}_j$ (several subsamples of each size could be considered).
3. Solve the ordinary least squares problem (or a robust analogue) given by $(\hat{\beta}_0, \hat{\beta}_1) = \underset{\beta_0, \beta_1}{\arg\min} \sum_{i=1}^{s} (\log(\hat{h}_i) - \beta_0 - \beta_1 \log(m_i))^2$, in which case $\hat{c} = e^{\hat{\beta}_0}$ and $\hat{p} = \hat{\beta}_1$ is our estimate of the order of convergence of the AMISE bandwidth.
4. Our estimate of the AMISE bandwidth for the original sample size, n, would be $\hat{h}_{AMISE,n} = \hat{c}n^{\hat{p}}$.

3. Simulation Study

Let us consider samples of size $n = 10^6$ drawn from the model $Y = m(X) + \varepsilon$, where $X \sim Beta(2,2)$, $\varepsilon \sim N(0, 0.2^2)$ and $m(x) = 1 + x\sin(5.5\pi x)^2$. Furthermore, we have considered a Gaussian kernel and, as a weight function, $w(x) = 1_{\{F_X^{-1}(0.05) \le x \le F_X^{-1}(0.95)\}}$, where F_X^{-1} denotes the marginal quantile function of X.

It is clear from Figure 1 that the OSS selector outperforms the SSS selector in terms of statistical precision. Moreover, in many cases bandwidths that are quite distant from the optimum do not have an associated large error (in terms of AMISE). On the other hand, as we can observe in Tables 1 and 2, the OSS selector is substantially faster than the SSS selector due to the fact that the former works with a single subsample size which, in turn, is even smaller than most of those considered for the SSS selector). It should be noted that the source code for both selectors was written in C++ and run in parallel on an Intel Core i5-8600K 3.6 GHz.

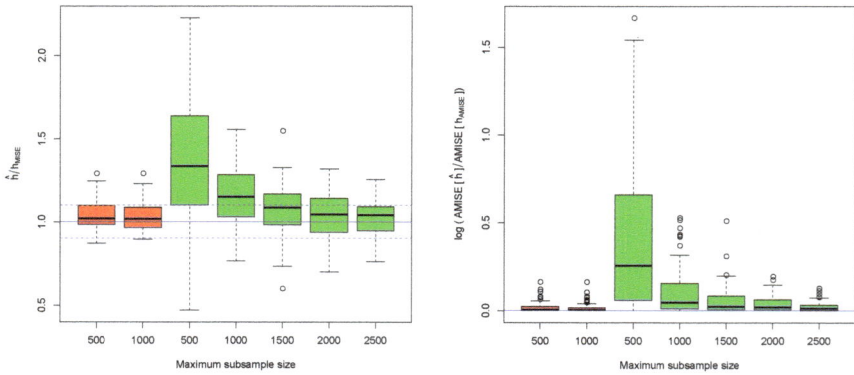

Figure 1. Sampling distributions of $\frac{\hat{h}}{h_{MISE,n}}$ (**left figure**) and $\log\left(\frac{AMISE(\hat{h})}{AMISE(h_{AMISE,n})}\right)$ (**right figure**) for the OSS (red) and SSS (green) bandwidth selectors.

Table 1. CPU elapsed times for the OSS selector with $n = 10^6$. 10 subsamples of the corresponding size were considered.

Subsample Size	CPU Elapsed Time (s)
500	1.62
1000	2.82

Table 2. CPU elapsed times for the SSS selector with $n = 10^6$ considering uniform grids (of 20 elements) of subsample sizes ranging from 100 to the corresponding maximum size. 10 subsamples of each of the corresponding sizes were considered.

Maximum Subsample Size	CPU Elapsed Time (s)
500	9.21
1000	17.9
1500	32.1
2000	51.0
2500	75.1

Author Contributions: Conceptualization, D.B., R.C. and M.F.; Methodology, D.B., R.C. and M.F.; Software, D.B., R.C. and M.F.; Validation, D.B., R.C. and M.F.; Formal Analysis, D.B., R.C. and M.F.; Investigation, D.B., R.C. and M.F.; Resources, D.B., R.C. and M.F.; Data Curation, D.B., R.C. and M.F.; Writing—Original Draft Preparation, D.B., R.C. and M.F.; Writing—Review & Editing, D.B., R.C. and M.F.; Visualization, D.B., R.C. and M.F.; Supervision, D.B., R.C. and M.F.; Project Administration, D.B., R.C. and M.F.; Funding Acquisition, D.B., R.C. and M.F.

Funding: This research received no external funding.

Acknowledgments: This research has been supported by MINECO grant MTM-2014-52876-R and by the Xunta de Galicia (Grupos de Referencia Competitiva ED431C-2016-015 and Centro Singular de Investigación de Galicia ED431G/01), all of them through the ERDF.

Conflicts of Interest: The authors declare no conflict of interest. The founding sponsors had no role in the design of the study; in the collection, analyses, or interpretation of data; in the writing of the manuscript, and in the decision to publish the results.

Proceedings **2018**, *2*, 1166

References

1. Nadaraya, E.A. On estimating regression. *Theory Probab. Its Appl.* **1964**, *9*, 141–142.
2. Wang, Q.; Lindsey, B.G. Improving cross-validated bandwidth selection using subsampling-extrapolation techniques. *Comput. Stat. Data Anal.* **2015**, *89*, 51–71.

proceedings

MDPI

Extended Abstract

Nonparametric Mean Estimation for Big-but-Biased Data [†]

Laura Borrajo * and Ricardo Cao

Research Group MODES, CITIC, Department of Mathematics, University of A Coruña, 15071 A Coruña, Spain; ricardo.cao@udc.es
* Correspondence: laura.borrajo@udc.es; Tel.: +34-981-167-000-1301
† Presented at the XoveTIC Congress, A Coruña, Spain, 27–28 September 2018.

Published: 19 September 2018

Abstract: Some authors have recently warned about the risks of the sentence *with enough data, the numbers speak for themselves*. The problem of nonparametric statistical inference in big data under the presence of sampling bias is considered in this work. The mean estimation problem is studied in this setup, in a nonparametric framework, when the biasing weight function is unknown (realistic). The problem of ignoring the weight function is remedied by having a small SRS of the real population. This problem is related to nonparametric density estimation. The asymptotic expression for the MSE of the estimator proposed is considered. Some simulations illustrate the performance of the nonparametric method proposed in this work.

Keywords: Bias Correction; Big Data; Kernel Method; mean estimation; Nonparametric Inference

1. Introduction

At certain times a large sample is not representative of the population, but it is biased (B3D). Some of the problems coming from ignoring sampling bias in big data statistical analysis have been recently reported by Cao [1]. A good example cited by Crawford [2] is the data collected in the city of Boston through the StreetBump smartphone app that underestimates the number of potholes in some neighborhoods of the city, with the consequent deficient management of resources. Another example is the database of more than 20 million tweets generated by Hurricane Sandy. These data come from a biased sample of the population, since most of the tweets came from Manhattan, while few tweets were originated in the most affected areas by the catastrophe. In other examples, such as those cited in Hargittai [3], survey data show that the use of sites is biased yielding samples that limit the generalizability of findings.

In this context, let us consider a population with CDF F (density f) and consider a SRS, $\mathbf{X} = (X_1, \ldots, X_n)$, of size n from this population. Assume that we are not able to observe this sample but we observe, instead, another sample $\mathbf{Y} = (Y_1, \ldots, Y_N)$, of a much larger sample size ($N >> n$) from a biased distribution G (density g), such that $g(x) = w(x)f(x)$, for some weight function $w(x) \geq 0$, $\forall x$.

2. Mean Estimation in B3D

To deal with the mean estimation problem in this context, we propose the realistic estimator (unknown w case) whose motivation is explained by Cao and Borrajo [4]:

$$\hat{\mu}^{\hat{w}_{h,b}} = \frac{\dfrac{1}{N}\sum_{i=1}^{N}\dfrac{Y_i}{\hat{w}_{h,b}(Y_i)}}{\dfrac{1}{N}\sum_{i=1}^{N}\dfrac{1}{\hat{w}_{h,b}(Y_i)}} = \frac{\dfrac{1}{N}\sum_{i=1}^{N}Y_i\dfrac{\hat{f}_h(Y_i)}{\hat{g}_b(Y_i)}}{\dfrac{1}{N}\sum_{i=1}^{N}\dfrac{\hat{f}_h(Y_i)}{\hat{g}_b(Y_i)}}. \tag{1}$$

In order to work with this estimator, extra information is required. We propose a scenario in which, in addition to the biased sample, **Y**, we also observe a SRS, **X**, of small size of the real population. The Parzen-Rosenblatt KDE (see [5,6]) based on **X** and **Y** can be used to estimate f and g.

The final expression of the AMSE of (1) ($h \to 0, b \to 0, nh \to \infty, Nb \to \infty$ and $N/n \to \infty$) is:

$$AMSE\left(\hat{\mu}^{\hat{w}_{h,b}}\right) = \left(C_1 b^2 + \frac{C_2}{Nb}\right)^2 + \frac{C_3}{n} + \frac{C_4}{Nn} + \frac{C_5}{N^2} + \frac{C_6}{Nnh} + \frac{C_7}{N^2 b}$$
$$+ \frac{C_8 h^2}{N^2 b} + \frac{C_9 h^4}{N} + \frac{C_{10} b^4}{N} + \frac{C_{11} h^2 b^2}{N} + \frac{C_{12} h}{Nn} + \frac{C_{13} b}{N^2}.$$

3. Case Study with Simulated Data

Let us consider $f(x) = \frac{3}{14}(x^2 + 1)\,\mathbf{1}_{[0,2]}(x)$ and $w(x) = 1.5\,\mathbf{1}_{[0,1.5]}(x) + x\,\mathbf{1}_{(1.5,2]}(x)$ (Figure 1a):

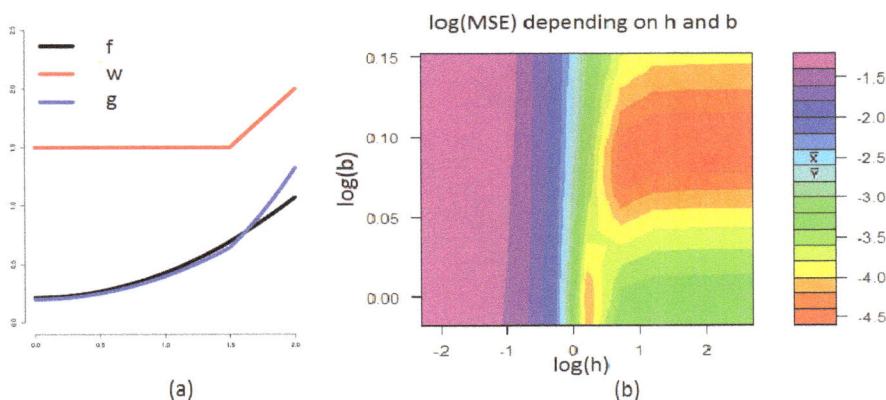

Figure 1. (**a**) Densities involved in the model. (**b**) Logarithm of the MSE of mu depending on the logarithm of h and b for this model, considering $n = 100$ and $N = 10,000$.

Figure 1b shows that the proposed estimator improves the estimation performed using the SRS, $\overline{X}$, and the biased sample, $\overline{Y}$, for a large number of combinations of h and b. Looking at Table 1, we observe that the best choice for h and b based on the simulation study contradicts the assumption ($h \to 0, b \to 0$) used in obtaining the asymptotic results. The AMSE for (1) under these non-standard asymptotic conditions ($h \to h_0, b \to b_0$) is:

$$AMSE\left(\hat{\mu}^{\hat{w}_{h_0,b_0}}\right) = \frac{D_1}{N} + \frac{D_2}{Nn} + \frac{D_3}{N^2} + \frac{D_4}{N^3}.$$

Table 1. MSE of the different estimators and optimal bandwidths obtained from the simulation study.

n	N	$MSE(\overline{X})$	$MSE(\overline{Y})$	$MSE(\hat{\mu}^{w_{h,b}})$	h	b
10	100	2.9×10^{-2}	4.4×10^{-3}	2.4×10^{-3}	1.99	1.05
50	2500	5.6×10^{-3}	1.7×10^{-3}	9.9×10^{-5}	3.97	1.18
100	10,000	2.9×10^{-3}	1.6×10^{-3}	2.5×10^{-5}	5.00	1.20
500	250,000	5.0×10^{-4}	1.6×10^{-3}	1.1×10^{-6}	12.22	1.23
1000	1,000,000	2.0×10^{-4}	1.6×10^{-3}	2.7×10^{-7}	12.22	1.24

4. Conclusions

Big Data brings new statistical challenges since bias is much more present. Ideas from length-biased data and nonparametric smoothing techniques are important in this context, testing for bias is a relevant problem in Big Data and smoothing parameter selection may be paradoxical in B3D.

Funding: This research has been supported by MINECO Grants MTM2014-52876-R and MTM2017-82724-R and by the Xunta de Galicia (Grupos de Referencia Competitiva ED431C-2016-015 and Centro Singular de Investigación de Galicia ED431G/01), all of them through the European Regional Development Fund (ERDF). The second author's research was sponsored by the Xunta de Galicia predoctoral grant (with reference ED481A-2016/367) for the universities of the Galician University System, public research organizations in Galicia and other entities of the Galician R&D&I System, whose funding comes from the European Social Fund (ESF) in 80% and in the remaining 20% from the General Secretary of Universities, belonging to the Ministry of Culture, Education and University Management of the Xunta de Galicia.

Conflicts of Interest: The authors declare no conflict of interest. The founding sponsors had no role in the design of the study; in the collection, analyses, or interpretation of data; in the writing of the manuscript, and in the decision to publish the results.

Abbreviations

The following abbreviations are used in this manuscript:

AMSE   Asymptotic mean squared error
B3D   Big-but-biased Data (BBBD)
CDF   Cumulative distribution function
KDE   Kernel density estimator
MSE   Mean squared error
SRS   Simple random sample

References

1. Cao, R. Inferencia estadística con datos de gran volumen. *Gac. RSME* **2015**, *18*, 393–417.
2. Crawford, K. The hidden biases in big data. *Harv. Bus. Rev.* **2013**. Available online: https://hbr.org/2013/04/the-hidden-biases-in-big-data (accessed on 4 April 2016).
3. Hargittai, E. Is Bigger Always Better? Potential Biases of Big Data Derived from Social Network Sites. *Ann. Am. Acad. Political Soc. Sci.* **2015**, *659*, 63–76.
4. Cao, R.; Borrajo, L. Nonparametric Mean Estimation for Big-But-Biased Data. In *The Mathematics of the Uncertain, Studies in Systems, Decision and Control*; Springer: Cham, Switzerland, 2018; Volume 142, pp. 55–65.
5. Parzen, E. On Estimation of a Probability Density Function and Mode. *Ann. Math. Stat.* **1962**, *33*, 1065–1076.
6. Rosenblatt, M. Remarks on some nonparametric estimates of a density function. *Ann. Math. Stat.* **1956**, *27*, 832–837.

proceedings

MDPI

Extended Abstract

Automatic System for the Identification and Visualization of the Retinal Vessel Tree Using OCT Imaging [†]

Joaquim de Moura [1,2,*], Jorge Novo [1,2], Noelia Barreira [1,2], Manuel G. Penedo [1,2] and Marcos Ortega [1,2]

[1] Department of Computing, University of A Coruña, 15071 A Coruña, Spain; jnovo@udc.es (J.N.); nbarreira@udc.es (N.B.); mgpenedo@udc.es (M.G.P.); mortega@udc.es (M.O.)

[2] CITIC-Research Center of Information and Communication Technologies, University of A Coruña, 15071 A Coruña, Spain

* Correspondence: joaquim.demoura@udc.es; Tel.: +34-981-167-000

† Presented at the XoveTIC Congress, A Coruña, Spain, 27–28 September 2018.

Published: 17 September 2018

Abstract: This paper proposes a system for the three-dimensional identification and visualization of the retinal vasculature using Optical Coherence Tomography (OCT) scans. This fully automatic tool provides useful biomarkers to the medical specialists that facilitate the prevention, diagnosis and treatment of various retinal and systemic pathologies.

Keywords: computer-aided diagnosis; Retinal imaging; Optical Coherence Tomography; vasculature; Retinal microcirculation

1. Introduction

The eye fundus is the only part of the human body where the blood vessels can be directly visualized non-invasively and safely in real time. An accurate analysis of the morphology of the retinal microvasculature allows the early diagnosis of different systemic diseases that can affect multiple parts of the body, such as hypertension, diabetes or arteriosclerosis, all of which are leading global public health concerns.

Nowadays, Optical Coherence Tomography (OCT) technique is increasing its use in clinical studies, health service research and daily clinical practice. This biomedical imaging system is capable of providing high-resolution cross-sectional scans of the internal microstructure of different retinal tissues in real time. These OCT scans allow the medical specialists to analyze and identify anatomical and physiological changes in the complex microvascular network of the retina.

In this work, we propose the development of a novel and complete methodology for the automatic identification and three-dimensional visualization of the retinal vasculature using OCT scans. This fully automatic system provides an intuitive visualization tool that facilitates a more complete and reliable analysis of the complex vascular structure of the retina.

2. Methodology

The proposed methodology is divided into four main steps: a first step, where we extract the two-dimensional vessel tree in the NIR retinography image; second, we estimate the vessel calibers; third, we obtain the corresponding depth of the vessel profiles in the OCT sections; and a final fourth step, where we make a complete three-dimensional representation of the retinal vasculature. This system allows the manipulation of the retinal vessel tree by means of graphical transformations including translation, scaling and rotation and their arbitrary combinations [1].

3. Results and Conclusions

The proposed method reached satisfactory results, providing a consistent and coherent three-dimensional retinal vessel tree reconstruction and visualization that can be posteriorly used in different medical analysis and diagnostic processes of various retinal and systemic pathologies. Figure 1 shows a representative example of the three-dimensional reconstruction and visualization where the caliber scale bar is also presented in the screen to facilitate the analysis of the medical specialists.

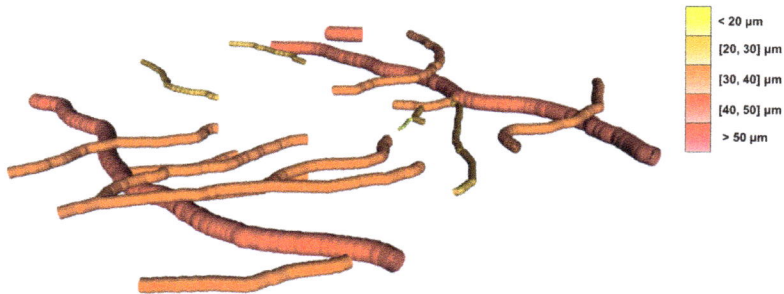

Figure 1. Representative example of the three-dimensional reconstruction and visualization of the retinal vessel tree using OCT images. Note the caliber scale bar that facilitates the analysis of the specialist.

Author Contributions: J.d.M., J.N. and N.B. contributed to the analysis and design of the computer methods and the experimental evaluation methods. M.G.P. and M.O. contributed with domain-specific knowledge. All the authors performed the result analysis. J.d.M. was in charge of writing the manuscript, and all the authors participated in its critical revision and final approval.

Acknowledgments: This work is supported by the Instituto de Salud Carlos III, Government of Spain and FEDER funds of the European Union through the PI14/02161 and the DTS15/00153 research projects and by the Ministerio de Economía y Competitividad, Government of Spain through the DPI2015-69948-R research project. Also, this work has received financial support from the European Union (European Regional Development Fund-ERDF) and the Xunta de Galicia, Centro singular de investigación de Galicia accreditation 2016-2019, Ref. ED431G/01; and Grupos de Referencia Competitiva, Ref. ED431C 2016-047.

Conflicts of Interest: The authors declare no conflict of interest. The founding sponsors had no role in the design of the study; in the collection, analyses, or interpretation of data; in the writing of the manuscript, and in the decision to publish the results.

References

1. De Moura, J.; Novo, J.; Charlón, P.; Barreira, N.; Ortega, M. Enhanced visualization of the retinal vasculature using depth information in OCT. *Med. Biol. Eng. Comput.* **2017**, *55*, 2209–2225.

proceedings

MDPI

Extended Abstract

Automatic Segmentation and Measurement of Vascular Biomarkers in OCT-A Images †

Macarena Díaz [1,2,*], Jorge Novo [1,2], Manuel G. Penedo [1,2] and Marcos Ortega [1,2]

[1] Department of Computation, University of A Coruña, 15071 A Coruña, Spain; jnovo@udc.es (J.N.); mgpenedo@udc.es (M.G.P.); mortega@udc.es (M.O.)

[2] CITIC-Research Center of Information and Communication Technologies, 15071 A Coruña, Spain

* Correspondence: macarena.diaz1@udc.es; Tel.: +34-981-167-000 (ext. 1330)

† Presented at the XoveTIC Congress, A Coruña, Spain, 27–28 September 2018.

Published: 17 September 2018

Abstract: We propose an automatic methodology that identifies the vascularity zones in OCT-A images and their measurement for its use in clinical analysis and diagnostic processes. The segmentation and measurement contributes objectivity and repeatability in the results, desirable characteristics in any diagnosis and monitoring process. In the validation of the method, the correlation coefficient of Pearson and Jaccard index were used, obtaining satisfactory results.

Keywords: computer-aided diagnosis; Image Segmentation; retinal imaging; Optical Coherence Tomography Angiography; vascularity

1. Introduction

Optical Coherence Tomography Angiography (OCT-A) is a new non-invasive imaging modality that facilitates the analysis of the vascularity in the retina. The extraction of this vascular and avascular zones is useful for the analysis of several pathologies such as diabetic retinopathy, but their correct extraction requires objectivity, determinism and repeatability factors. Given the recent appearance of this image modality, there are few works, most of them are clinical proposals that study the repeatability and reproducibility of different biomarkers that are based on the OCT-A vascular properties in healthy patients, indicating the satisfactory impact of this analysis. For this reason, the automatic extraction of this zones is interesting, given the repeatability and objectivity that support its automation.

2. Methodology

We propose an automatic methodology that identifies the vascular and avascular zones in OCT-A images and their measurement for its use in clinical analysis and diagnostic processes [1]. We firstly intensify the vascular characteristics using morphological operators, facilitating the extraction in further steps. Then, a set of image processing techniques are combined to maximize their differences and, posteriorly, estimate their representative parameters, respectively. These biomarkers are based in the area of the Foveal Avascular Zone (FAZ) and the vascular density, features that can vary in healthy and pathological patients. In the case of the vascular density, four different ways of measurement were performed based on: the original image, the enhanced image, the thresholded image and the vascular skeletonization of the image. Figure 1, represents an example of input and outputs system.

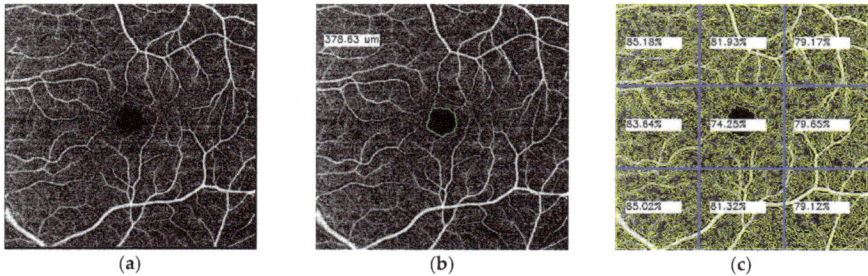

Figure 1. Example of the input (**a**) of the system and the outputs: avascular zone (**b**) and vascular zone (**c**).

3. Results

The proposed methodology was tested on a set of 144 non-pathological images labeled by an expert ophtalmologist, being used as reference in the validation of the method. The correlation coefficient of Pearson and the Jaccard index were used to validate the results between the expert and the system, validating the measurements and the coverage of the zone. In the correlation coefficient of Pearson, the areas of the expert and the areas of the system were compared, obtaining an average of 0.76, which represents a good correlation between both segmentations. With the Jaccard index, we obtained 0.73, also offering satisfactory results. Summarizing, the proposed methodology presented satisfactory results in both validation experiments.

Author Contributions: M.D. and J.N. contributed to the analysis and design of the automatic methods and the experimental evaluation methods. M.G.P. and M.O. contributed with domain-specific knowledge. All the authors performed the result analysis. M.D. was in charge of writing the manuscript, and all the authors participated in the revision and final approval.

Acknowledgments: This work is supported by the Instituto de Salud Carlos III, Government of Spain and FEDER funds of the European Union through the PI14/02161 and the DTS15/00153 research projects and by the Ministerio de Economía y Competitividad, Government of Spain through the DPI2015-69948-R research project. Also, this work has received nancial support from the European Union (European Regional Development Fund - ERDF) and the Xunta de Galicia, Centro singular de investigación de Galicia accreditation 2016–2019, Ref. ED431G/01; and Grupos de Referencia Competitiva, Ref. ED431C 2016-047.

Conflicts of Interest: The authors declare no conflict of interest. The founding sponsors had no role in the design of the study; in the collection, analyses, or interpretation of data; in the writing of the manuscript, and in the decision to publish the results.

References

1. Díaz, M.; Novo, J.; Penedo, M.G.; Ortega, M. Automatic extraction of vascularity measurements using OCT-A images. KES'18 - International Conference on Knowledge Based and Intelligent Information and Engineering Systems, Belgrado, Serbia, September 2018, Procedia Computer Science, 126, 273-281.

proceedings

MDPI

Extended Abstract

On the Processing and Analysis of Microtexts: From Normalization to Semantics [†]

Yerai Doval [1,*] and David Vilares [2]

[1] Grupo COLE, Departamento de Informática, Escuela Superior de Ingeniería Informática, Universidade de Vigo, Campus As Lagoas, 32004 Ourense, Spain

[2] FASTPARSE Lab, Grupo LyS, Departamento de Computación, Facultade de Informática, Universidade da Coruña, Campus de Elviña, 15071 A Coruña, Spain; david.vilares@udc.es

* Correspondence: yerai.doval@uvigo.es; Tel.: +34-988-387-280

[†] Presented at the XoveTIC Congress, A Coruña, Spain, 27–28 September 2018.

Published: 18 September 2018

Abstract: User-generated content published on microblogging social platforms constitutes an invaluable source of information for diverse purposes: health surveillance, business intelligence, political analysis, etc. We present an overview of our work on the field of microtext processing covering the entire pipeline: from input preprocessing to high-level text mining applications.

Keywords: microtext normalization; Language Identification; sentiment analysis; text preprocessing; text mining; semantics

1. Introduction

Extracting information from microtexts (e.g., tweets) requires the use of Natural Language Processing (NLP) techniques. Unfortunately, their performance is sensitive to the so-called texting phenomena (shortenings, substitutions, word concatenation, etc.) present in these texts. Thus, we first need to adapt the input to writing standards in a process called microtext normalization.

2. Microtext Normalization

One of the most usual approaches when implementing a microtext normalization system is decomposing it into two steps [1]: normalization candidate generation, where domain dictionaries, phonetic algorithms [2], as well as other spell checking techniques are used to obtain standard words to replace in the input text; and candidate selection, where the most likely normalized sequence according to some language model is constructed.

Notably, this approach works at the word level, as candidates are generated and selected for each word in the input text. However, word boundaries (in this case, blank spaces) are also affected by texting phenomena, hence their positioning cannot be assumed to be correct.

To address this issue we can add, as an early step in the normalization pipeline, a word segmentation subsystem that will try to normalize the positioning of word boundaries. In particular, we have experimented with character-based n-gram language models paired with a beam search algorithm, obtaining state-of-the-art results [3].

On top of this, in order to support multilingual environments such as most microblogging social platforms, it becomes essential to know in advance the language or languages in which the texts we want to normalize are written in, so that we can choose the right modules for the task. Consequently, we have added an automatic language identifier to our normalization pipeline. In this regard, we have tested and adapted well-known tools for the task [4].

The ongoing work is currently focusing on obtaining an accurate candidate selection mechanism, where language models play again a key role.

3. Sentiment Analysis

Normalization systems have many applications in downstream NLP tasks, such as Sentiment Analysis (SA) in Twitter, where the goal is to predict the polarity of a text being positive, negative or neutral. In this context, we have studied symbolic systems that compute the sentiment of sentences by taking into account their syntactic structure. The hypothesis is that syntactic relations between pairs of words are helpful to process linguistic phenomena such as negation, intensification or adversative subordinate clauses, very relevant for the task at hand. Our experiments suggest that our approach better deals with these phenomena than lexical-based systems. We also have developed machine learning models that have been evaluated in international evaluation campaigns [5,6].

These techniques are usually applied to monolingual environments, but their application to multilingual and code-switching texts, where words coming from two or more languages are used indistinctly, is gaining increasing interest [7].

Normalization and sentiment analysis might also be useful in higher level text mining applications. Political analysis, where the main goal is to use social media to estimate the popularity of politicians, is of special interest as it can be used as an alternative to traditional polls [8].

Furthermore, NLP techniques can be used in social analysis to study the cultural differences across different countries. More in particular, in [9] we explore the semantics of part-of-day nouns for different cultures in Twitter, which can be helpful to understand how different societies organize their day schedule.

Author Contributions: Y.D. conceived, designed and performed the normalization experiments; Y.D. analyzed the data from the normalization experiments; D.V. conceived, designed and performed the sentiment analysis experiments; D.V. analyzed the data from the sentiment analysis experiments; Y.D. and D.V. wrote the paper.

Acknowledgments: Research partially funded by the Spanish Ministry of Economy, Industry and Competitiveness (MINECO) through projects FFI2014-51978-C2-2-R, TIN2017–85160–C2–1–R and TIN2017–85160–C2–2–R; the Spanish State Secretariat for Research, Development and Innovation (which belongs to MINECO) and the European Social Fund (ESF) under a FPI fellowship (BES-2015-073768) associated to project FFI2014-51978-C2-1-R; and by the Galician Regional Government under project ED431D 2017/12. This research has received funding from the European Research Council (ERC) under the European Union's Horizon 2020 research and innovation programme (grant agreement No. 714150 – FASTPARSE), which covered the costs of open access publishing. We gratefully acknowledge NVIDIA Corporation for the donation of a GTX Titan X GPU used for this research.

Conflicts of Interest: The authors declare no conflict of interest. The founding sponsors had no role in the design of the study; in the collection, analyses, or interpretation of data; in the writing of the manuscript, and in the decision to publish the results.

References

1. Doval, Y.; Vilares, J.; Gómez-Rodríguez, C. LYSGROUP: Adapting a Spanish microtext normalization system to English. In Proceedings of the Workshop on Noisy User-generated Text, Beijing, China, 31 July 2015; pp. 99–105.
2. Doval, Y.; Vilares, M.; Vilares, J. On the performance of phonetic algorithms in microtext normalization. *ESWA* **2018**, *113*, 213–222.
3. Doval, Y.; Gómez-Rodríguez, C. Comparing Neural- and N-gram-based Language Models for Word Segmentation. *JASIST* **2018**, accepted.

4. Doval, Y.; Vilares, D.; Vilares, J. Identificación Automática del Idioma en Twitter: Adaptación de Identificadores del Estado del Arte al Contexto Ibérico. In Proceedings of the Tweet Language Identification Workshop co-located with the 30th Conference of the Spanish Society for Natural Language Processing (SEPLN 2014), Girona, Spain, 16 September 2014; Zubiaga, A., Vicente, I.S., Gamallo, P., Pichel, J.R., Alegria, I., Aranberri, N., Ezeiza, A., Fresno, V., Eds.; CEUR Workshop Proceedings, 2014; Volume 1228, pp. 39–43.

5. Vilares, D.; Doval, Y.; Alonso, M.A.; Gómez-Rodríguez, C. LYS at SemEval-2016 Task 4: Exploiting neural activation values for Twitter sentiment classification and quantification. In Proceedings of the 10th International Workshop on Semantic Evaluation, San Diego, CA, USA, 16–17 June 2016; pp. 79–84.

6. Vilares, D.; Doval, Y.; Alonso, M.A.; Gómez-Rodríguez, C. LyS at TASS 2015: Deep Learning Experiments for Sentiment Analysis on Spanish Tweets. In Proceedings of the Workshop on Sentiment Analysis at SEPLN, Alicante, Spain, 15 September 2015; Villena-Román, J., García-Morera, J., García-Cumbreras, M.A., Martínez-Cámara, E., Martín-Valdivia, M.T., Ureña-López, L.A., Eds.; CEUR Workshop Proceedings, 2015; Volume 1397, pp. 47–52.

7. Vilares, D.; Alonso, M.A.; Gómez-Rodríguez, C. Supervised sentiment analysis in multilingual environments. *IPM* **2017**, *53*, 595–607.

8. Vilares, D.; Thelwall, M.; Alonso, M.A. The megaphone of the people? Spanish SentiStrength for real-time analysis of political tweets. *JIS* **2015**, *41*, 799–813.

9. Vilares, D.; Gómez-Rodríguez, C. Grounding the Semantics of Part-of-Day Nouns Worldwide using Twitter. In Proceedings of the 2nd Workshop on Computational Modeling of People's Opinions, Personality, and Emotions in Social Media, New Orleans, Louisiana, 6 June 2018; pp. 123–128.

proceedings

MDPI

Extended Abstract

Interpretable Market Segmentation on High Dimension Data [†]

Carlos Eiras-Franco [1,*], Bertha Guijarro-Berdiñas [1], Amparo Alonso-Betanzos [1] and Antonio Bahamonde [2]

[1] Grupo LIDIA, CITIC, Universidade da Coruña, 15071 A Coruña, Spain; berta.guijarro@udc.es (B.G.-B.); amparo.alonso.betanzos@udc.es (A.A.-B.)

[2] Computer Science Department, Universidad de Oviedo, 33203 Gijón, Spain; abahamonde@uniovi.es

* Correspondence: carlos.eiras.franco@udc.es

[†] Presented at the XoveTIC Congress. A Coruña, Spain, 27–28 September 2018.

Published: 17 September 2018

Abstract: Obtaining relevant information from the vast amount of data generated by interactions in a market or, in general, from a dyadic dataset, is a broad problem of great interest both for industry and academia. Also, the interpretability of machine learning algorithms is becoming increasingly relevant and even becoming a legal requirement, all of which increases the demand for such algorithms. In this work we propose a quality measure that factors in the interpretability of results. Additionally, we present a grouping algorithm on dyadic data that returns results with a level of interpretability selected by the user and capable of handling large volumes of data. Experiments show the accuracy of the results, on par with traditional methods, as well as its scalability.

Keywords: market segmentation; interpretability; Explainability; scalability; Machine Learning; Big Data

1. Introduction

Data obtained by monitoring a marketplace are mainly dyadic [1], that is, they represent the relation between two entities (for instance user vs products, buyers vs sellers or any other pairing of agents). This sort of data are also prevalent in common problems such as recommender systems [2], computational linguistics, information retrieval and preference learning [3], besides being used in more specific problems like automatic test grading [4].

A traditional problem to be solved with this kind of data consists on obtaining groups of entities that show a similar behavior. Market segmentation is the process of performing this analysis on market data [5]. The resulting grouping is coveted by companies since it offers valuable insight, but it is hard to obtain.

Also, having results that are easily interpretable by managers is essential. Interpretability is given by a collection of characteristics that promote ease of understanding of a model [6] and can be achieved by providing transparent models and algorithms or by offering additional explanations for the outputs of the model.

The algorithm introduced in this work aims to obtain informative and easy to interpret data for human supervisors. It is implemented in the Apache Spark [7] distributed framework, which enables the analysis of large amounts of data in a reasonable time.

2. Proposal

Given a dataset $\mathcal{X}$ containing data showing the interactions between two entities $\mathcal{U}$ and $\mathcal{V}$ in which each data point $x \in \mathcal{X}$ has the form $(u, i, f(u, i))$ with f being a *utility function* $f : (\mathcal{U}, \mathcal{I}) \rightarrow \{-1, +1\}$, a grouping $Cl(\mathcal{U}) = \{Clu_1, \ldots, Clu_m\}$ on one of the entities can be defined as a set of m groups containing all the elements in $\mathcal{U}$. The aptness of this grouping can be measured as the homogeneity of the value v across the elements in each Clu_k [8]. Using this measure, we can define the *weighted entropy* of a grouping as

$$WE(Cl(\mathcal{U})) = \sum_{k,j} \frac{|Clu_k|}{|\mathcal{U}||\mathcal{I}|} H\left(\frac{|\{u \in Clu_k : f(u, i_j) = +1\}|}{|Clu_k|}\right). \tag{1}$$

where $H(x)$ represents the Shannon entropy of x.

Since each $u \in \mathcal{U}$ is a defined by a set of variables, each Clu_k is defined by giving a range for those variables. We can obtain a measure of the *quality* of $Cl(\mathcal{U})$ by adding a factor that measures its interpretability. We do that by adding the number of such variables needed to define each group Clu_k.

$$Q(Cl(\mathcal{U})) = -WE(Cl(\mathcal{U})) - \lambda \sum_{Clu_k \in Cl(\mathcal{U})} NV(Clu_k). \tag{2}$$

where $NV(x)$ represents the number of variables needed to characterize x and λ is a hyperparameter that enables the user to manage the balance between accuracy and ease of understanding.

The proposed algorithm takes a dataset $\mathcal{X}$ as input and returns a grouping $Cl(\mathcal{U})$ that maximizes $Q(Cl(\mathcal{U}))$. It does so by constructing a decision tree over the variables in $\mathcal{U}$ which defines the grouping $Cl(\mathcal{U})$.

Algorithm 1: Grouping algorithm.

Data: $\mathcal{U}, f, L_{MAX}$(MAX DEPTH OF THE TREE), N(MEASURES THE EXPLORATION SPACE)
Result: Decision tree that defines the grouping.
function BUILDTREE($\mathcal{U}, level, splitPoints, f, L_{MAX}, N$)

    if *level* $>L\_MAX$ **then**
        return $\varnothing$
    end
    candidates $\leftarrow$ sorted list with capacity N;
    for $(variable, value) \in splitPoints$ **do**
        1   *left* $\leftarrow \{u \in \mathcal{U} : u[variable] < value\}$;
                right $\leftarrow \{u \in \mathcal{U} : u[variable] > value\}$;
                if HEURISTIC(*left, right*) $>$ *candidates.minimum* **then**
        2   *candidates.add*((*variable, value*));
                end
    end
    best $\leftarrow \varnothing$;
    for $(variable, value) \in candidates$ **do**
        3   *left* $\leftarrow \{u \in \mathcal{U} : u[variable] < value\}$;
        4   *right* $\leftarrow \{u \in \mathcal{U} : u[variable] > value\}$;
        5   *treeLeft* $\leftarrow$ BUILDTREE(*left, level* + 1, *splitPoints*);
        6   *treeRight* $\leftarrow$ BUILDTREE(*right, level* + 1, *splitPoints*);
        7   *new* $\leftarrow$ (*variable, value, leftTree, rightTree*) **if** WE(*new, f*) $>$ WE(*best, f*) **then**
        8   *best* = *new*;
                end
    end
    return *new*;
end
splitPoints $\leftarrow$ list of split points in every variable;
return BUILDTREE($\mathcal{U}, 0, splitPoints, f, L_{MAX}, N$);

---

3. Results

Experiments performed on a large real world dataset containing information about readers and news items show that the proposed algorithm obtains a grouping consisting of 18 groups with a weighted entropy similar to that of the grouping with 100 elements obtained by Kmeans with $k = 100$.

Additional experiments show that the Apache Spark implementation of the algorithm shows almost linear scalability when adding more computation nodes.

4. Acknowledgements

This work has been partially funded by the Ministerio de Economía y Competitividad (research projects TIN 2015-65069-C2, both 1-R and 2-R and "Red Española de Big Data y Análisis de datos escalable", TIN2016-82013-REDT), by the Xunta de Galicia (GRC2014/035 y ED431G/01) and by the European Union Regional Development Funds.

The authors want to thank Fundación Pública Galega Centro Tecnolóxico de Supercomputación de Galicia (CESGA) for the use of their computing resources.

Conflicts of Interest: The authors declare no conflict of interest. The founding sponsors had no role in the design of the study; in the collection, analyses, or interpretation of data; in the writing of the manuscript, and in the decision to publish the results.

References

1. Hofmann, T.; Puzicha, J.; Jordan, M.I. Learning from dyadic data. In *Advances in Neural Information Processing Systems*; MIT Press: Cambridge, MA, USA, 1999; pp. 466–472.
2. Koren, Y.; Bell, R.; Volinsky, C. Matrix factorization techniques for recommender systems. *Computer* **2009**, *42*, 30–37.
3. Luaces, O.; Díez, J.; Alonso-Betanzos, A.; Troncoso, A.; Bahamonde, A. A factorization approach to evaluate open-response assignments in MOOCs using preference learning on peer assessments. *Knowl. Based Syst.* **2015**, *85*, 322–328.
4. Luaces, O.; Díez, J.; Alonso-Betanzos, A.; Troncoso, A.; Bahamonde, A. Content-based methods in peer assessment of open-response questions to grade students as authors and as graders. *Knowl. Based Syst.* **2017**, *117*, 79–87.
5. Kotler, P.; Cox, K.K. *Marketing Management and Strategy*; Prentice Hall: Upper Saddle River, NJ, USA, 1980.
6. Lipton, Z.C. The mythos of model interpretability. *arXiv* **2016**, arXiv:1606.03490.
7. Zaharia, M.; Xin, R.S.; Wendell, P.; Das, T.; Armbrust, M.; Dave, A.; Meng, X.; Rosen, J.; Venkataraman, S.; Franklin, M.J.; et al. Apache spark: A unified engine for big data processing. *Commun. ACM* **2016**, *59*, 56–65.
8. Díez, J.; Pérez, P.; Luaces, O.; Bahamonde, A. *Readers Segmentation According to their Preferences to Click Promoted Links in Digital Publications*; Technical Report; Universidad de Oviedo, Oviedo, Spain, 2018.

proceedings

MDPI

Extended Abstract

Reconstruction of Tomographic Images through Machine Learning Techniques †

Xosé Fernández-Fuentes [1,*,‡], David Mera [1,‡] and Andrés Gómez [2,‡]

1   Centro Singular de Investigación en Tecnoloxías da Información (CiTIUS), Universidade de Santiago de Compostela, Rúa de Jenaro de la Fuente Domínguez, 15782 Santiago de Compostela, Spain; david.mera@usc.es

2   Galicia Supercomputing Center (CESGA), 15705 Santiago de Compostela, Spain; agomez@cesga.es

*   Correspondence: xosefernandez.fuentes@usc.es; Tel.: +34-881-816-390

†   Presented at the XoveTIC Congress, A Coruña, Spain, 27–28 September 2018.

‡   These authors contributed equally to this work.

Published: 17 September 2018

check for updates

Abstract: Some problems in the field of health or industry require to obtain information from the inside of a body without using invasive methods. Some techniques are able to get qualitative images. However, these images are not enough to solve some problems that require an accurate knowledge. Normally, the tomography processes are used to explore inside of a body. In this particular case, we are using the method called Electrical Impedance Tomography (EIT). The basic operation of this method is as follows: (1) The electrical potential difference is measured in the electrodes placed around the body. This part is known as forward model. (2) Get information from the inside of the body using the measured voltages. This problem is known as inverse problem. There are several approximations to solve this inverse problem. However, these solutions are focused on obtaining qualitative images. In this paper, we show the main challenges of how to obtain quantitative knowledge when Machine Learning techniques are used to solve this inverse problem.

Keywords: Electrical Impedance Tomography; Machine Learning; Artificial Neural Networks; inverse problems

1. Introduction

Certain medical and industrial problems need to get information from the inside of a body without damaging it. There are techniques able to get qualitative images about the distribution of some physical characteristic of a particular body. However, these images are not useful to solve problems that require a quantitative knowledge of a concrete physical feature. One of the ways to get this knowledge is using tomography processes, which explore the inside of a body in a non-invasive way. In this paper, we have focused our efforts on the tomography process called Electrical Impedance Tomography (EIT). We have chosen EIT because it can be useful for both medicine and industry due to the fact it is easy to deploy.

Signals obtained through a tomography process are used to solve an ill-posed [1,2] nonlinear inverse problem with the purpose of obtaining the distribution of a physical characteristic. To deal with this mathematical problem exist different approaches. On the one hand, some techniques apply iterative algorithms [3,4]. They are quite accurate but they demand a lot of time and large computing capacity. On the other hand, there are some algorithms that assume some linearity in the response of the body. These algorithms are fast but inaccurate [5,6].

In this paper, we show what are the main challenges to solve this inverse problem in an accurate and fast way thanks to Machine Learning (ML) techniques. Some previous work [7–9] explore the possibility of using ML to solve the inverse problem. However, their focus is on reconstructing the images from a qualitative point of view instead of quantitative.

2. Challenges

The first problem is to get an appropriate dataset. Because there is not a dataset large enough to train the models, it is necessary to generate it. This new dataset must contain simulations of bodies with distributions and volumes of different physical properties. In addition, it is necessary to take into account the position and dimensions of the electrodes used. Once this set has been generated, it is necessary to simulate the tomographic processes to each body. In the case of EIT, the software EIDORS [10] is appropriate to perform this simulation. After the simulation is done, noise should be added to make the signals more similar to those obtained in a real environment. This implies studying and defining an appropriate function that allows the signals to be distorted in an appropriate way.

The second challenge is to train different algorithms of Machine Learning, with the objective of making a comparison that allows identifying the algorithms that best adapt to this type of highly non-linear problem. It must be taken into account that the final algorithm must be robust to the noise of the sensors, the body shape and the position of the sensors. Furthermore, to be able to train so many different models, it is necessary to have available a large computing capacity and a lot of storage. At the same time, this implies to carry out an efficient management of the computational resources.

The third challenge (very related to the previous one) is to develop adequate metrics for the training of the algorithms. Traditional metrics do not seem appropriate due to the atypical unbalance between inputs and outputs (many more outputs than inputs).

The fourth problem is the validation of the final algorithm. This implies testing the model with a real test set that has been obtained through tomographic tests.

3. Results

We have done some initial tests using Artificial Neural Networks. To perform these first tests, we have made various simplifications. For example, we have maintained the same body shape and we have not introduced noise in the measurements of the sensors. Having this in mind, the results obtained are very interesting and promising taking into account the complex nature of the problem.

Acknowledgments: This work has received financial support from the Predoctoral scholarship program of the Xunta de Galicia (ED481A-2018/277), the Xunta de Galicia under Research Network R2016/045, the Consellería de Cultura, Educación e Ordenación Universitaria (accreditation 2016-2019, ED431G/08) and the European Regional Development Fund (ERDF). Computational resources were provided by the Galicia Supercomputing Center (CESGA).

Conflicts of Interest: The authors declare no conflict of interest. The founding sponsors had no role in the design of the study; in the collection, analyses, or interpretation of data; in the writing of the manuscript, and in the decision to publish the results.

References

1. Michalikova, M.; Abed, R.; Prauzek, M.; Koziorek, J. Image Reconstruction in Electrical Impedance Tomography Using Neural Network. In Proceedings of the Biomedical Engineering Conference (CIBEC), Giza, Egypt, 11–13 December 2014; pp. 39–42.
2. Wang, P.; Li, H.L.; Xie, L.L.; Sun, Y.C. The implementation of FEM and RBF neural network in EIT. In Proceedings of the 2nd International Conference on Intelligent Networks and Intelligent Systems (ICINIS 2009), Tianjin, China, 1–3 November 2008; Volume 3, pp. 66–69.
3. Wang, C.; Lang, L.; Wang, H.-X. RBF neural network image reconstruction for electrical impedance tomography. In Proceedings of the 2004 International Conference on Machine Learning and Cybernetics, Shanghai, China, 26–29 August 2004; Volume 4, pp. 2549–2552.
4. Adler, A.; Guardo, R. A Neural Network Image Reconstruction Technique for Electrical Impedance Tomography. *IEEE Trans. Med. Imaging* 1994, 13, 594–600.

5. Wu, K.; Yang, J.; Dong, X.; Fu, F.; Tao, F.; Liu, S. Comparative study of reconstruction algorithms for electrical impedance tomography. *IEEE Trans. Biomed. Eng.* **2012**, *51077127*, 2296–2299.

6. Guardo, R.; Boulay, C.; Murray, B.; Bertrand, M. An experimental study in electrical impedance tomography using backprojection reconstruction. *IEEE Trans. Biomed. Eng.* **1991**, *38*, 617–627.

7. Liu, X.; Wang, X.; Hu, H.; Li, L.; Yang, X. An extreme learning machine combined with Landweber iteration algorithm for the inverse problem of electrical capacitance tomography. *Flow Meas. Instrum.* **2015**, *45*, 348–356.

8. Martin, S.; Choi, C.T. Nonlinear Electrical Impedance Tomography Reconstruction Using Artificial Neural Networks and Particle Swarm Optimization. *IEEE Trans. Magn.* **2016**, *52*, 1–4.

9. Martin, S.; Choi, C.T.M. A Post-Processing Method for Three-Dimensional Electrical Impedance Tomography. *Sci. Rep.* **2017**, *7*, 7212.

10. Adler, A.; Lionheart, W.R. Uses and abuses of EIDORS: An extensible software base for EIT. *Physiol. Meas.* **2006**, *27*, S25.

proceedings

MDPI

Extended Abstract

Network Data Unsupervised Clustering to Anomaly Detection [†]

Manuel López-Vizcaíno [iD] *, Carlos Dafonte [iD], Francisco J. Nóvoa [iD], Daniel Garabato [iD] and M. A. Álvarez [iD]

CITIC, UDC, Campus de Elviña s/n, 15071 A Coruña, Spain; carlos.dafonte@udc.es (C.D.); francisco.javier.novoa@udc.es (F.J.N.); daniel.garabato@udc.es (D.G.); marco.antonio.agonzalez@udc.es (M.A.Á.)
* Correspondence: manuel.fernandezl@udc.es
† Presented at the XoveTIC Congress, A Coruña, Spain, 27–28 September 2018.

Published: 17 September 2018

check for updates

Abstract: In these days, organizations rely on the availability and security of their communication networks to perform daily operations. As a result, network data must be analyzed in order to provide an adequate level of security and to detect anomalies or malfunctions in the systems. Due to the increase of devices connected to these networks, the complexity to analyze data related to its communications also grows. We propose a method, based on Self-Organized Maps, which combine numerical and categorical features, to ease communication network data analysis. Also, we have explored the possibility of using different sources of data.

Keywords: Self-Organizing Maps; IDS; network security; categorical SOM; visualization; unsupervised clustering

1. Introduction

These days network data analysis has become essential to provide adequate levels of security in mid and big sized networks. The number of connected devices has increased to 20 thousand million of devices in 2017 and will exceed 30 thousand million devices in 2020, as it is reflected in the forecast from Statista [1]. As a result of the exponential increase of the traffic generated, classical analysis techniques based on payload packet inspection become unfeasible [2].

One possible approximation to this problem is unsupervised clustering. These techniques allow to cluster elements with similar characteristics without prior knowledge, easing the analysis of the data as it was shown in previous research [3]. In particular for the scope of this work we have chosen Self-Organized Maps (SOM) technique [4] as it allows to perform clustering as well as dimensionality reduction. Also, this technique has been successfully applied to Intrusion Detection Systems [5,6].

As it is said in [7], a habitual traffic profile, called baseline, is present in communication networks. Different kind of attacks present deviations from this baseline and these features could be used to detect certain anomalies in traffic behavior (DoS [8], DDoS [9], brute force attacks [10]).

The objective of this work is to present a method to ease the analysis of communication network data. Providing a system to allow the study and detection of anomalies out of data gathered from different sources.

2. Methods

For the scope of this work, to generate the clusters, we have modified SOM technique to accept numerical and categorical features, as explained in [11]. Besides we have only used information present on IP packet headers or values derived from them. From the data available on the IP header we have selected a number of features such as source, destination, source port, destination port, protocol, duration and bytes transmitted.

Two different datasets have been used to perform the experiments. One the one hand, the UNB ISCX which is a synthetic and labeled flow dataset generated by the Cybersecurity Institute of Canada intended for Intrusion Detection research [12]. On the other hand, we have used a log dataset gathered in the firewall of the Computer Science faculty of the University of A Coruña. We have divided both datasets to use the 80% of them to train and the 20% were left to test.

Before the clustering technique could be applied, a preprocessing step must be preformed in order to categorize certain variables and to normalize numeric features. We performed a shallow approach to the analysis of the map configuration with three different map sizes: 10×10 (100 neurons), 20×20 (400 neurons) and 30×30 (900 neurons). Increasing the number of neurons could help to get better detection rates but it also rises the complexity of map analysis.

Finally, to evaluate the clustering we have used some tags referred to the nature of the connection. In the case of the flow dataset we have used the synthetic labels showing if it is part of an attack or normal traffic. On the other dataset we have taken the actions of the firewall as an approach. In the last case it also allows the revision of the firewall rules by studying the misclassification.

3. Results

The aim of these experiments was to determine if a mixed numerical-categorical version of SOM technique was suitable for network data classification, by using only IP header information gathered from different sources. Also, other objective was to study how the information obtained could be used in relation with the source of the data. For example, network flows could help to detect attacks and firewall logs analysis could help to detect misconfiguration.

As it can bee seen in Table 1 where the results are shown both for the flow labeled dataset (ISCX) and the firewall log (FIC), there are similarities between their results. Bigger map sizes tend to increase the performance of the technique for both datasets but with the drawback of a more complex map analysis. Also, it should be noticed that the better overall results are achieved with logs rather than with flows.

Table 1. Experiment results.

	Flows			Logs		
	10×10	20×20	30×30	10×10	20×20	30×30
Sensitivity	90.33%	94.09%	94.28%	87.78%	90.20%	94.66%
Specificity	98.36%	99.00%	99.26%	96.37%	99.24%	99.12%
Precision	67.06%	77.80%	82.44%	86.56%	96.95%	96.62%
Accuracy	98.07%	98.83%	99.08%	94.56%	97.34%	98.18%

4. Discussion

As it can be seen in the results, despite the differences between both datasets, we can conclude that the technique could be applied to different sources of network data. This difference should be studied in order to determine if it is related to the nature of the dataset, the differences in the features or other reasons. Also, additional research using other sources of data and different configurations over the proposed technique should be performed.

Funding: This work has received financial support from the Xunta de Galicia (Centro singular de investigación de Galicia accreditation 2016-2019) and the European Union (European Regional Development Fund - ERDF).

Conflicts of Interest: The authors declare no conflict of interest. The founding sponsors had no role in the design of the study; in the collection, analyses, or interpretation of data; in the writing of the manuscript, and in the decision to publish the results.

References

1. Statista. IHS. Internet of Things (Iot) Connected Devices Installed Base Worldwide from 2015 to 2025 (in Billions) 2018. Available online: https://www.statista.com/statistics/471264/iot-number-of-connected-devices-worldwide/ (accessed on 17 September 2018).
2. Umer, M.F.; Sher, M.; Bi, Y. Flow-based intrusion detection: Techniques and challenges. *Comput. Secur.* **2017**, *70*, 238–254, doi:10.1016/j.cose.2017.05.009.
3. Buczak, A.; Guven, E. A survey of data mining and machine learning methods for cyber security intrusion detection. *IEEE Commun. Surv. Tutor.* **2016**, *18*, 1153–1176, doi:10.1109/COMST.2015.2494502.
4. Kohonen, T. Self-organized formation of topologically correct feature maps. *Biol. Cybern.* **1982**, *43*, 59–69, doi:10.1007/BF00337288.
5. Ibrahim, L.M.; Basheer, D.T.; Mahmod, M.S. A comparison study for intrusion database (KDD99, NSL-KDD) based on self organization map (SOM) artificial neural network. *J. Eng. Sci. Technol.* **2013**, *8*, 107–119.
6. Ramadas, M.; Ostermann, S.; Tjaden, B. Detecting Anomalous Network Traffic with Self-organizing Maps. In *Recent Advances in Intrusion Detection*; Vigna, G., Kruegel, C., Jonsson, E., Eds.; Springer: Berlin/Heidelberg, Germany, 2003; pp. 36–54.
7. Xu, K.; Zhang, Z.L.; Bhattacharyya, S. Internet Traffic Behavior Profiling for Network Security Monitoring. *IEEE ACM Trans. Netw.* **2008**, *16*, 1241–1252, doi:10.1109/TNET.2007.911438.
8. Fadlullah, Z.M.; Taleb, T.; Vasilakos, A.V.; Guizani, M.; Kato, N. DTRAB: Combating Against Attacks on Encrypted Protocols Through Traffic-Feature Analysis. *IEEE ACM Trans. Netw.* **2010**, *18*, 1234–1247, doi:10.1109/TNET.2009.2039492.
9. Lee, K.; Kim, J.; Kwon, K.H.; Han, Y.; Kim, S. DDoS attack detection method using cluster analysis. *Expert Syst. Appl.* **2008**, *34*, 1659–1665, doi:10.1016/j.eswa.2007.01.040.
10. Hofstede, R.; Jonker, M.; Sperotto, A.; Pras, A. Flow-Based Web Application Brute-Force Attack and Compromise Detection. *J. Netw. Syst. Manag.* **2017**, *25*, 735–758, doi:10.1007/s10922-017-9421-4.
11. Del Coso, C.; Fustes, D.; Dafonte, C.; Nóvoa, F.J.; Rodríguez-Pedreira, J.M.; Arcay, B. Mixing numerical and categorical data in a Self-Organizing Map by means of frequency neurons. *Appl. Soft Comput.* **2015**, *36*, 246–254, doi:10.1016/J.ASOC.2015.06.058.
12. Shiravi, A.; Shiravi, H.; Tavallaee, M.; Ghorbani, A.A. Toward developing a systematic approach to generate benchmark datasets for intrusion detection. *Comput. Secur.* **2012**, *31*, 357–374, doi:10.1016/j.cose.2011.12.012.

proceedings

Extended Abstract

A Convolutional Network for the Classification of Sleep Stages †

Isaac Fernández-Varela ⬤, Elena Hernández-Pereira and Vicente Moret-Bonillo

CITIC, Universidade da Coruña, 15071 A Coruña, Spain; elena@udc.es (E.H.-P.); vicente.moret@udc.es (V.M.-B.)
* Correspondence: isaac.fvarela@udc.es
† Presented at the XoveTIC Congress, A Coruña, Spain, 27–28 September 2018.

Published: 14 September 2018

check for updates

Abstract: The classification of sleep stages is a crucial task in the context of sleep medicine. It involves the analysis of multiple signals thus being tedious and complex. Even for a trained physician scoring a whole night sleep study can take several hours. Most of the automatic methods trying to solve this problem use human engineered features biased for a specific dataset. In this work we use deep learning to avoid human bias. We propose an ensemble of 5 convolutional networks achieving a kappa index of 0.83 when classifying 500 sleep studies.

Keywords: sleep staging; convolutional neural network; classification

1. Introduction

Sleep disorders are a common problem: insomnia has a prevalence of 20% and daytime sleepiness between 12% and 15% [1,2]. Sleep disorders can be diagnosed analysing a set of bio-signals recorded during the sleep period, a technique called polysomnography. This analysis is expensive, uncomfortable for the patient and difficult to interpret. Thus, it is usually presented as an hypnogram, a graph showing the evolution of the sleep stages.

The gold standard for the hypnogram construction is the American Academy of Sleep Medicine (AASM) guide, which includes how to identify sleep stages and associated events such as arousals, movements and cardiac and respiratory events. This guide identifies 5 sleep stages: Awake (W), Rapid Eye Movements (REM), and 3 Non REM known as N1, N2, and N3. A well built hypnogram allows a quickly and accurate diagnosis. Yet, the agreement between two experts trying to build the same hypnogram is lower than 90% (with a kappa index between 0.48 and 0.89 [3]), with even less agreement for specific stages such as N1.

The aforementioned reasons motivated several works that automate the sleep stages classification. Traditionally, these works were based on feature extraction and later classification [4–8], solutions commonly biased towards the available dataset. To solve the bias problem we propose the use of Deep Learning, an option already explored by some authors [9–13].

Particularly, we use a convolutional network that learns the relevant features for the classification by itself. Following the AASM guides we use multiple channels, namely two electroencephalogram (EEG), one electromyogram (EMG) and both electrooculogram (EOG). Furthermore, our signals are filtered to reduce noise and remove artefacts induced by the electrocardiogram (ECG).

2. Materials

Our experiments were carried out using real polisomnographs (PSG) from the Sleep Heart Health Study (SHHS) [14]. These PSG were scored by several experts following the AASM rules [15] and include 2 EEG, both EOG, EMG and ECG.

From the database we randomly selected 3 datasets for training, testing and validation containing 400, 100, and 500 registers respectively, or 288,000, 119,121, and 606,981 samples. Most of the samples belong to class N2 (36%) or W (38%), and the less represented one is N1 (3%). Imbalanced classes is a typical problem in sleep medicine.

3. Method

We use a convolutional network that is fed with 5 filtered signals simultaneously: two EEG derivations, both EOG and one EMG. The filtering pipeline includes a Notch filter in 60 Hz for all the signals, and a high pass in 15 Hz for the EMG signal. We also remove ECG artefacts using an adaptive filter [16].

Following clinical procedure, the network input are 30 s windows (usually called epochs) with the signals re-sampled (if needed) to 125 Hz, resulting in a sample dimension of 3750 × 5. Each signal is normalised to mean 0 and deviation 1, using as reference the training dataset.

Figure 1 represents the proposed convolutional network. The convolutional block presented in the figure is a set of four layers including: 1D convolution, batch normalization, ReLu activation and average pool. This block is repeated n times. All the 1D convolutions have the same kernel size but layer i has twice the filters of layer $i - 1$.

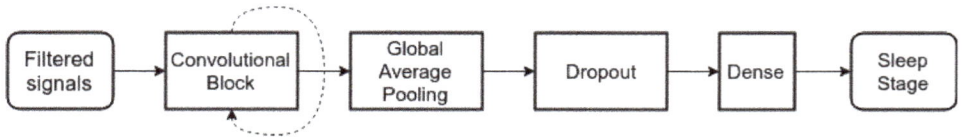

Figure 1. Proposed Convolutional Network.

The network was trained using Adam optimiser with 64 samples per batch and early stopping with a patience of 10, monitoring the loss made in the validation dataset.

To select the hyper-parameters: n, the number of filters for the first layer, kernel size, and learning rate; we used a Tree-structured Parzen Estimator (TPE), which is sequential model-based optimisation (SMBO) approach. We trained 50 models using the TPE and selected the 5 best to build an ensemble.

4. Results

The ensemble built with the 5 best models was used to carry out experiments with the test dataset, obtaining the performance measures and confusion matrix shown in Figure 2. The class with the best classification is W and then N2, N3, REM show similar values regarding the F1 score, although there are significant differences in the sensitivity. As expected, N1 is the class with the worst classification, with values lower than 0.4. Apart from the problems classifying class N1, most of the errors happen between classes N2 and N3.

Stage	Precision	Sensitivity	F1 score
W	0.94	0.96	0.95
N1	0.39	0.21	0.27
N2	0.87	0.89	0.88
N3	0.92	0.77	0.84
REM	0.82	0.90	0.86
Average	0.78	0.75	0.76

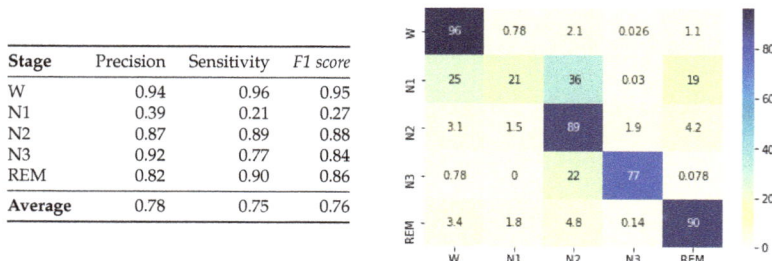

Figure 2. Confusion matrix for the test set classification using the 5 model ensemble.

5. Discussion and Conclusions

In this work we propose an ensemble of convolutional networks to classify sleep stages. The main reason is to avoid introducing human bias in our solution with a method that learns the relevant features by itself.

To configure the network hyper-parameters we used a tree-structured parzen estimator, evaluating 50 different models and selecting the best 5 to build an ensemble. This ensemble achieves an average precision, sensitivity, and F1 score of 0.78, 0.75 and 0.76 with a kappa index of 0.83. Yet it shows difficulties to classify class N1 and a bias towards class N2.

Given the lack of standards or benchmarks related to this problem, it is difficult to compare our solutions against previous works. Some references are shown in Table 1. Our values are competitive, achieving the highest kappa index and the best classification for class W.

Table 1. Performance achieved in previous works.

Work	Dataset	Kappa	F1 Score				
			W	N1	N2	N3	REM
Biswal et al. [12]	Massachusetts General Hospital, 1000 recordings	0.77	0.81	**0.70**	0.77	0.83	**0.92**
Längkvist et al. [9]	St Vicent's University Hospital, 25 recordings	0.63	0.73	0.44	0.65	**0.86**	0.80
Sors et al. [13]	SHHS, 1730 recordings	0.81	0.91	0.43	0.88	0.85	0.85
Supratak et al. [11]	MASS dataset, 62 recordings	0,80	0,87	0,60	**0.90**	0.82	0.89
Supratak et al. [11]	SleepEDF, 20 recordings	0.76	0.85	0.47	0.86	0.85	0.82
Tsinalis et al. [10]	SleepEDF, 39 recordings	0.71	0.72	0.47	0.85	0.84	0.81
Tsinalis et al.[17]	SleepEDF, 39 recordings	0.66	0.67	0.44	0.81	0.85	0.76
This work	SHHS, 500 recordings	**0.83**	**0.95**	0.27	0.88	0.84	0.86

Results are promising and the approach should be easily extended to other PSG sources. Moreover, if we could have PSG acquired in different conditions during the training phase, regularisation should improve.

As future work, we need to understand how and why the model makes the decisions instead of treating it as a black box.

Funding: This research was partially financed by the Xunta de Galicia [ED431G/01] and the European Union through the ERDF fund.

Acknowledgments: We gratefully acknowledge the support of NVIDIA Corporation with the donation of the Titan Xp GPU used for this research.

Conflicts of Interest: The authors declare no conflict of interest. The founding sponsors had no role in the design of the study; in the collection, analyses, or interpretation of data; in the writing of the manuscript, and in the decision to publish the results.

References

1. Ohayon, M.M.; Sagales, T. Prevalence of insomnia and sleep characteristics in the general population of Spain. *Sleep Med.* **2010**, *11*, 1010–1018, doi:10.1016/j.sleep.2010.02.018.

2. Marin, J.; Gascon, J.M.; Carrizo, S.; Gispert, J. Prevalence of sleep apnoea syndrome in the Spanish adult population. *Int. J. Epidemiol.* **1997**, *26*, 381–386, doi:10.1093/ije/26.2.381.

3. Stepnowsky, C.; Levendowski, D.; Popovic, D.; Ayappa, I.; Rapoport, D.M. Scoring accuracy of automated sleep staging from a bipolar electroocular recording compared to manual scoring by multiple raters. *Sleep Med.* **2013**, *14*, 1199–1207, doi:10.1016/j.sleep.2013.04.022.

4. Liang, J.; Lu, R.; Zhang, C.; Wang, F. Predicting Seizures from Electroencephalography Recordings: A Knowledge Transfer Strategy. In Proceedings of the 2016 IEEE International Conference on Healthcare Informatics (ICHI 2016), Chicago, IL, USA, 4–7 October 2016; pp. 184–191, doi:10.1109/ICHI.2016.27.

5. Hassan, A.R.; Bhuiyan, M.I.H. A decision support system for automatic sleep staging from EEG signals using tunable Q-factor wavelet transform and spectral features. *J. Neurosci. Meth.* **2016**, *271*, 107–118, doi:10.1016/J.JNEUMETH.2016.07.012.

6. Sharma, R.; Pachori, R.B.; Upadhyay, A. Automatic sleep stages classification based on iterative filtering of electroencephalogram signals. *Neural Comput. Appl.* **2017**, *28*, 2959–2978, doi:10.1007/s00521-017-2919-6.

7. Lajnef, T.; Chaibi, S.; Ruby, P.; Aguera, P.E.; Eichenlaub, J.B.; Samet, M.; Kachouri, A.; Jerbi, K. Learning machines and sleeping brains: Automatic sleep stage classification using decision-tree multi-class support vector machines. *J. Neurosci. Meth.* **2015**, *250*, 94–105, doi:10.1016/J.JNEUMETH.2015.01.022.

8. Huang, C.S.; Lin, C.L.; Ko, L.W.; Liu, S.Y.; Su, T.P.; Lin, C.T. Knowledge-based identification of sleep stages based on two forehead electroencephalogram channels. *Front. Neurosci.* **2014**, *8*, 263, doi:10.3389/fnins.2014.00263.

9. Längkvist, M.; Karlsson, L.; Loutfi, A. Sleep Stage Classification Using Unsupervised Feature Learning. *Adv. Artif. Neural Syst.* **2012**, *2012*, 1–9, doi:10.1155/2012/107046.

10. Tsinalis, O.; Matthews, P.M.; Guo, Y.; Zafeiriou, S. Automatic Sleep Stage Scoring with Single-Channel EEG Using Convolutional Neural Networks. *arXiv* **2016**, arXiv:1610.01683.

11. Supratak, A.; Dong, H.; Wu, C.; Guo, Y. DeepSleepNet: A Model for Automatic Sleep Stage Scoring Based on Raw Single-Channel EEG. *IEEE Trans. Neural Syst. Rehabil. Eng.* **2017**, *25*, 1998–2008, doi:10.1109/TNSRE.2017.2721116.

12. Biswal, S.; Kulas, J.; Sun, H.; Goparaju, B.; Westover, M.B.; Bianchi, M.T.; Sun, J. SLEEPNET: Automated Sleep Staging System via Deep Learning. *arXiv* **2017**, arXiv:1707.08262.

13. Sors, A.; Bonnet, S.; Mirek, S.; Vercueil, L.; Payen, J.F. A convolutional neural network for sleep stage scoring from raw single-channel EEG. *Biomed. Signal Process. Control* **2018**, *42*, 107–114, doi:10.1016/j.bspc.2017.12.001.

14. Quan, S.F.; Howard, B.V.; Iber, C.; Kiley, J.P.; Nieto, F.J.; O'connor, G.T.; Rapoport, D.M.; Redline, S.; Robbins, J.; Samet, J.M.; et al. The Sleep Heart Health Study: Design, Rationale, and Methods. *Sleep* **1997**, *20*, 1077–1085, doi:10.1093/sleep/20.12.1077.

15. Bonnet, M.H.; Carley, D.; Carskadon, M. EEG arousals: Scoring rules and examples: a preliminary report from the Sleep Disorders Atlas Task Force of the American Sleep Disorders Association. *Sleep* **1992**, *15*, 173–184.

16. Fernández-Varela, I.; Alvarez-Estevez, D.; Hernández-Pereira, E.; Moret-Bonillo, V. A simple and robust method for the automatic scoring of EEG arousals in polysomnographic recordings. *Comput. Biol. Med.* **2017**, *87*, 77–86, doi:10.1016/J.COMPBIOMED.2017.05.011.

17. Tsinalis, O.; Matthews, P.M.; Guo, Y. Automatic Sleep Stage Scoring Using Time-Frequency Analysis and Stacked Sparse Autoencoders. *Ann. Biomed. Eng.* **2016**, *44*, 1587–1597, doi:10.1007/s10439-015-1444-y.

proceedings

MDPI

Extended Abstract

Analysis of the Effect of Tidal Level on the Discharge Capacity of Two Urban Rivers Using Bidimensional Numerical Modelling [†]

Ignacio Fraga

CITIC—Research Center of Information and Communication Technologies, University of A Coruña, 15071 A Coruña, Spain; ignacio.fraga@udc.es; Tel.: +34-981-167-000

† Presented at the XoveTIC Congress. A Coruña, Spain, 27–28 September 2018.

Published: 14 September 2018

Abstract: This article quantifies the variation of the discharge capacity of an urban river of the Galicia region due to the tidal level at the river discharge. During high tides, the water level on the river outlet produces a backwater effects that reduces the maximum discharge. This results in a decrease of the maximum capacity to one third of the maximum discharge during low tide.

Keywords: urban hydraulics; computer fluid dynamics; urban flood

1. Introduction

Floods are the most frequent natural disaster. Over the last two decades they have affected more than 2.500 million people and caused economic damages worth 625 billion dollars. Urban environments are becoming more sensitive to flood events due to the increase of the catchments imperviousness and the anthropic pressure on waterways.

This fact becomes significant for the urban waterways of the Galicia region. This region, located at the Northwest of Spain, has a very dense river network characterized by short rivers, many of them with important localities situated at the river discharge. Urban floods in Galicia are conditioned by both the amount of rainfall conveyed by the river and the tidal level at the discharge. During high tides, the sea level at the river outlet causes a backwater effect which reduces the waterway capacity. In this article, a study to quantify the effect of tidal level on the river discharges of two urban river reaches is presented.

2. Study Case

The studied rivers correspond to the Mendo and the Mandeo rivers, which comprise draining areas of 353 and 84 km² respectively. The lengths of the rivers are 50 km (Mandeo) and 30 km (Mendo) and both of them run along predominantly forest and agricultural lands. Due to its location at the Atlantic coast, the studied catchments are located on the path of low pressure fronts, resulting in frequent adverse storm events characterized by a large spatial and temporal variability of rainfall. The locality of Betanzos (around 30,000 inhabitants) is located at the junction of the Mendo and Mandeo rivers and suffers frequent floods.

3. Methodology

The effect of the tidal level on the maximum discharge capacity of the Mendo and the Mandeo rivers was analyzed using the Iber model [1]. This bidimensional flow model was used to simulate the hydrodynamic of the rivers on their pass through the locality of Betanzos. The model extended from approximately 1 km upstream the locality to the discharge of the Mandeo river at the Betanzos estuary. The domain was discretized in an unstructured mesh with elements size ranging between 60 m (at the estuary) to 1.5 m (at the river reaches along the locality of Betanzos). The elevation of each element was interpolated from the Digital Elevation Model provided by the Instituto Geográfico Nacional (IGN) with a spatial resolution of 5 m.

An inflow was imposed at each boundary condition of the Mendo and Mandeo rivers, increasing from 0 to 400 m³/s during 36 h for the Mandeo river, and from 0 to 34 m³/s during the same 36 h for the Mendo river. At the outlet boundary, a constant water level was imposed. Multiple simulations were performed with different levels in order to reproduce the tidal levels reported by the National Port Authorities of Spain for the Betanzos estuary. The maximum discharge was considered as the one that caused the river to overflow the natural waterway, as illustrated in Figure 1.

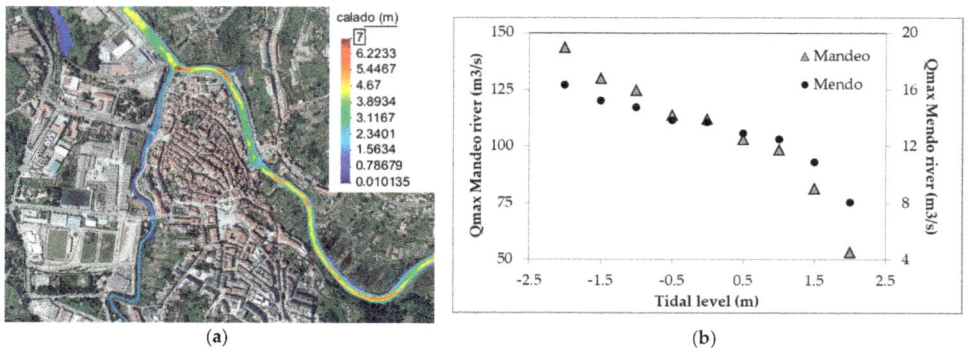

(a) (b)

Figure 1. (a) Example of water depth fields before flow exceeds the waterway (b) Maximum discharges obtained for the simulated tidal levels.

4. Results

Results show a significant decrease of the maximum discharge with the increase of tidal level (Figure 1). During the peak of spring tides, maximum discharge is nearly one third of the one corresponding to low tide in both rivers. Therefore, flood studies should be performed taking into account the impact of tidal level on the flood extent and the maximum discharge conveyed by the rivers.

Author Contributions: I.F. was responsible of the presented study, analysis and publication of the results.

Acknowledgments: This work received financial support from the Xunta de Galicia (Centro singular de investigación de Galicia accreditation 2016–2019) and the European Union (European Regional Development Fund-ERDF).

Conflicts of Interest: The author declares no conflict of interest. The founding sponsors had no role in the design of the study; in the collection, analyses, or interpretation of data; in the writing of the manuscript, and in the decision to publish the results.

References

1. Bladé, E.; Cea, L.; Corestein, G.; Escolano, E.; Puertas, J.; Vázquez-Cendón, E.; Dolz, J.; Coll, A. Iber: herramienta de simulación numérica del flujo en ríos. *Rev. Int. Metod. Numer.* **2014**, *30*, 1–10.

proceedings

MDPI

Extended Abstract

Channel Covariance Identification in FDD Massive MIMO Systems †

José P. González-Coma [1,*,‡] , **Pedro Suárez-Casal** [2,‡], **Paula M. Castro** [2,‡] **and Luis Castedo** [2,‡]

1   University of A Coruña, CITIC, A Coruña, Spain
2   Department of Computer Engineering, University of A Coruña, A Coruña, Spain;
    pedro.scasal@udc.es (P.S.-C.); paula.castro@udc.es (P.M.C.); luis.castedo@udc.es (L.C.)
*   Correspondence: jose.gcoma@udc.es; Tel.: +34-881-016-051
†   Presented at the XoveTIC Congress, A Coruña, Spain, 27–28 September 2018.
‡   These authors contributed equally to this work.

Published: 19 September 2018

check for
updates

Abstract: Channel estimation for Massive MIMO systems has drawn a lot of attention in the last years. A number of estimation methods rely on the knowledge of the channel covariance matrix to operate. However, this covariance is not known in practice, and it should be acquired. In this work, we investigate different techniques for covariance identification under the assumption of very short training sequences.

Keywords: covariance identification; Massive MIMO; FDD

1. Introduction

Due to the large data rates necessary for future communication systems, Massive Multiple-Input Multiple-Output (MIMO) constitutes a promising candidate radio technology [1]. One of the features of this technology is the use of large antenna arrays. Therefore, large channel matrices have to be estimated at the user end in Frequency-Division Duplex (FDD) mode. Specifically, channel estimation is usually achieved by transmitting training sequences known to the transmitter and the receiver whose length, on the one hand, must be sufficient to obtain accurate estimates and, on the other, should be as short as possible to not impair data transmission efficiency. These circumstances make channel estimation a difficult task.

In recent literature, some authors consider covariance identification based on the assumption of channel reciprocity between the downlink and the uplink [2]. Moreover, [3,4] consider the FDD case by employing angular reciprocity instead.

2. Materials and Methods

In this work, we will focus on channel covariance identification for FDD on a more general scenario, that is, we assume that the channel is a stationary process whose statistics remain constant during a certain period of time. In addition, typical antenna arrangements like Uniform Linear Arrays (ULA) lead to covariance matrices with Toeplitz structure.

MUltiple SIgnal Classification (MUSIC) is a well-known Angle of Departure (AoD) estimation algorithm. For the considered problem, the MUSIC algorithm identifies the angles corresponding to the channel propagation paths. Thanks to the Toeplitz structure of the covariance matrix, we circumvent the typical limitations of MUSIC when the associated channel gain variances have to be estimated. Further, when the rank of the sample covariance matrix is smaller than the number of propagation paths, an interesting scenario arises. To address channel covariance identification in this scenario, we resort to the technique Spatial smoothing by employing training sequences designed as sparse rulers [5]. We have improved the technique proposed in [6] by using all the available information.

3. Results

In this section, we present some numerical results obtained with the proposed algorithms, together with the comparison with previous methods in the literature.

The following setup is considered for the numerical experiments. The number of transmit antennas is 400 for the channel covariance model of a ULA, and different number of channel paths. The training sequence was generated as a sparse ruler of length 50. The methods evaluated are as follows:

- Covariance Orthogonal Matching Pursuit (COMP) [7]
- MUSIC
- Spatial Smoothing (SS)
- Improved Spatial Smoothing (ISS)
- Maximum Likelihood (ML) [8]

The performance metric is the Normalized Mean Squared Error (NMSE) between the actual channel covariance matrix and the estimated one. We try different levels of SNR, namely 0 dB and 30 dB. The number of training periods indicates the number of channel block employed to estimate the channel covariance matrix, and might be smaller than the number of channel propagation paths.

NMSE vs. training periods for a moderate number of channel paths is shown in Figure 1a. Remarkably, MUSIC is more robust than COMP for the low SNR regime. ML is the best strategy in terms of performance but the algorithm is also the most computationally expensive, and may be not practical for realistic scenarios. Moreover, the other methods converge to ML when the number of snapshots is large enough. Regarding SS and ISS curves, we illustrate the gain of our approach with respect to the standard one.

In Figure 1b the number of paths to estimate is much larger. MUSIC and COMP do not apply to this scenario due to the large number of parameters to be estimated. On the one hand, the restrictions of MUSIC regarding the sample covariance rank do not hold. On the other hand, COMP assumes sparsity, which does not apply in this setup. Therefore, we compare SS and ISS with ML that can be interpreted as a benchmark in the high SNR regime.

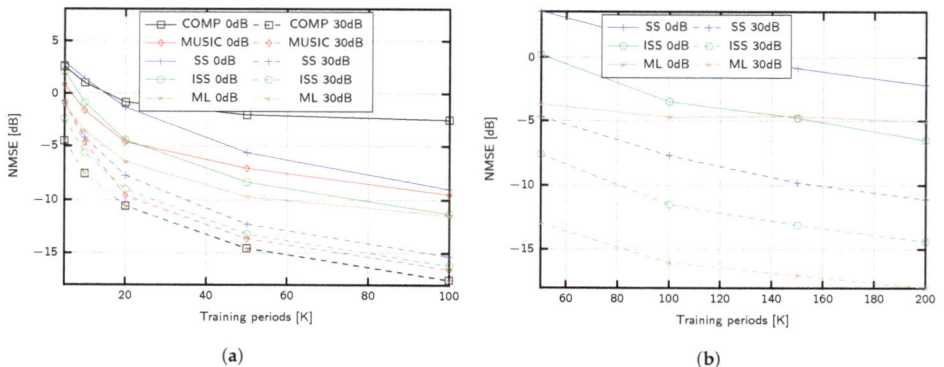

(a)   (b)

Figure 1. (a) NMSE of different covariance identification strategies for 400 antennas and 15 channel propagation paths. (b) NMSE of different covariance identification strategies for 400 antennas and 70 channel propagation paths.

4. Conclusions

We have analyzed methods to identify the covariance matrix even if the sample covariance matrix is rank deficient. Moreover, we have updated the spatial smoothing method to improve the estimation quality in terms of NMSE.

Funding: This work has been funded by Xunta de Galicia (ED431C 2016- 045, ED341D R2016/012, ED431G/01), AEI of Spain (TEC2015-69648-REDC, TEC2016-75067-C4-1-R), and ERDF funds (AEI/FEDER, EU).

Conflicts of Interest: The authors declare no conflict of interest. The founding sponsors had no role in the design of the study; in the collection, analyses, or interpretation of data; in the writing of the manuscript, and in the decision to publish the results.

References

1. Rusek, F.; Persson, D.; Lau, B.K.; Larsson, E.G.; Marzetta, T.L.; Edfors, O.; Tufvesson, F. Scaling up MIMO: Opportunities and Challenges with Very Large Arrays. *IEEE Signal Process. Mag.* **2013**, *30*, 40–60, doi:10.1109/MSP.2011.2178495.
2. Neumann, D.; Joham, M.; Utschick, W. Covariance Matrix Estimation in Massive MIMO. *IEEE Signal Process. Lett.* **2018**, *25*, 863–867, doi:10.1109/LSP.2018.2827323.
3. Xie, H.; Gao, F.; Jin, S.; Fang, J.; Liang, Y. Channel Estimation for TDD/FDD Massive MIMO Systems with Channel Covariance Computing. *arXiv* **2017**, arXiv:1710.00704.
4. Khalilsarai, M.B.; Haghighatshoar, S.; Yi, X.; Caire, G. FDD Massive MIMO via UL/DL Channel Covariance Extrapolation and Active Channel Sparsification. *arXiv* **2018**, arXiv:1803.05754.
5. Romero, D.; Leus, G. Compressive covariance sampling. In Proceedings of the 2013 Information Theory and Applications Workshop (ITA), San Diego, CA, USA, 10–15 February 2013; pp. 1–8, doi:10.1109/ITA.2013.6502949.
6. Ariananda, D.D.; Leus, G. Direction of arrival estimation for more correlated sources than active sensors. *Signal Process.* **2013**, *93*, 3435–3448, doi:10.1016/j.sigpro.2013.04.011.
7. Park, S.; Heath, R.W., Jr. Spatial Channel Covariance Estimation for the Hybrid MIMO Architecture: A Compressive Sensing Based Approach. *arXiv* **2017**, arXiv:1711.04207
8. Babu, P.; Stoica, P. Sparse spectral-line estimation for nonuniformly sampled multivariate time series: SPICE, LIKES and MSBL. In Proceedings of the 20th European Signal Process Conference (EUSIPCO), Bucharest, Romania, 27–31 August 2012; pp. 445–449.

proceedings

MDPI

Extended Abstract

Guidelines to Support Graphical User Interface Design for Children with Autism Spectrum Disorder: An Interdisciplinary Approach [†]

Betania Groba [1,2,*], Nereida Canosa [1,2] and Patricia Concheiro-Moscoso [1,2,]

[1] Research group Artificial Neural Networks and Adaptive Systems-Centre of Medical Informatics and Radiological Diagnosis (RNASA-IMEDIR), Research Center on Information and Communication Technologies (CITIC), Faculty of Health Sciences, Universidade da Coruña, 15071 A Coruña, Spain; nereida.canosa@udc.es (N.C.); patricia.concheiro@udc.es (P.C.-M.)
[2] Institute for Biomedical Research (INIBIC), Universidade da Coruña, 15071 A Coruña, Spain
* Correspondence: b.groba@udc.es
† Presented at the XoveTIC Congress, A Coruña, Spain, 27–28 September 2018.

Published: 17 September 2018

Abstract: The study aims to describe the guidelines to support user interface design for develop technology centered in the specific learning style, abilities and needs of children with Autism Spectrum Disorder (ASD). This research study describes the conclusions drawn following a process of interactive design of software, ASD Module, In-TIC PC. Four groups of participants were involved in the process: specialists with experience in the intervention with people with ASD, specialists with experience in the development and design of technology for people with disability, children with ASD and their families ($n = 39$). The techniques used to formalize the collection of information from different groups of participants were observation, interview, group discussions and a questionnaire. The results of the study target the development of a design guide that includes the evidence, the basic ideas and suggestions deduced from the design and development process of the ASD Module. This translates into a list of rules with suggestions to consider in the design and adaptation of technology for children with ASD. These guidelines of interface design provide useful information for researchers, developers, social and healthcare professionals and families, with the aim of offering alternatives for children with ASD and facilitating the understanding of daily life.

Keywords: Human-Computer Interface; Interdisciplinary Projects; Pedagogical Issues; Activities of Daily Living; Autism; children

1. Introduction

The type of technology used in interventions with people with ASD is varied: computers, mobile devices, video recordings, robots and virtual reality. In the last two decades, software developed in this field has increased in numbers [1]. In this regard, a clear example is the free or low-cost initiatives that offer technological solutions for people with ASD.

Technology developers expressed a clear interest to design programs that met the needs of this population [2–4]. Therefore, at present, not much evidence describes the developmental and design procedure that allows generating technology centered in the specific learning style, abilities and needs of children with ASD.

2. Material and Methods

A cross-sectional study design was employed. The Autonomous Ethics Committee of Research in Galicia approved the protocol (code: 2014/558).

2.1. Participants and Settings

Four groups of respondents were participated in the study (*n* = 39): Professionals with experience in the intervention with people with ASD (first group); Professionals with experience in the development and design of technology for people with disability (second group); Family members of people with ASD (third group) and children with ASD (fourth group).

2.2. Procedure

Research has been based on user-centered design and has followed an iterative procedure. This is a cyclical process, divided into the following phases:

- Study and analysis of the recent scientific evidence on the design of technology for people with ASD. Participants: 2nd Group.
- Study and analysis of the recent scientific evidence on the skills and ways of processing information by people with ASD. Participants: 2nd Group.
- Observation, analysis and discussion on the skills and ways of processing of this population and their influence on the design of technology. Participants: 1st, 2nd, 3rd, and 4th Groups.
- Design and development of the application. Participants: 2nd Group.
- Software testing by professionals and family members of people with ASD. Participants: 1st and 3rd Groups.
- Software quality improvement. Participants: 2nd Group.
- After this iterative process, the resulting application was tested by the 4th Group, that is, children with ASD.

3. Results

The result of the study targets the development of a design guide that includes the evidence, the basic ideas and suggestions deduced from the design and development process of the ASD Module. This translates into a list of rules with suggestions to consider in the customization and adaptation of technology for people with ASD. The rules extracted from the process are listed below:

1. The Software and its Contents Are Based on a Person's Abilities, Desires and Interests
2. The Design of the Interface is Simple and the Information Displayed is Simplified
3. Use of Images to Display Information
4. The Images Convey the Meaning of the Actual Element
5. The Use of Images Allows Users to Adapt According to their Level of Visual Cognition
6. The Image is Accompanied by the Written Word
7. Speech Synthesis is Used to Facilitate Communication or as Reinforcement to the Command
8. The Information is Displayed in a Multimodal Way (Visual and Auditory) and it is Adapted According to the Sensory Style Preferred by each Child
9. The Background Color is Used to Facilitate the Information Processing
10. The User has the Possibility to Customize all Relevant Aspects

4. Discussion

Some of the most complex decisions are related to the differences in the perception of people with ASD. In connection with this, it is explained that people with ASD process visual information more easily [3,5–9], although it is generally assumed that it is increasingly necessary to evaluate the sensory style of each person. Similarly, it is understood that the simultaneous display of auditory and visual stimuli help to learn, but in some cases, people with ASD need to be provided with information through their preferred sensory channel [3,10].

Finally, one of the premises supported by almost all the literature is the need for customization of technology.

5. Conclusions

This list of rules for technology design and customization provides useful information for researchers, developers, social and healthcare professionals and families, with the aim of offering alternatives for children with ASD and facilitating the understanding of daily life.

Author Contributions: B.G. and, N.C. conceived and designed the experiment; B.G. performed the experiment; N.C. and P.C.-M. analyzed the data; B.G., N.C. and P.C.-M. wrote the paper.

Acknowledgments: We would like to thank participants in the project for their contributions, knowledge and time. We would also like to thank families, professionals and children of aÚPa (children's and baby's clinic of Psychology and Physiotherapy of A Coruña). This study was supported by the Orange Foundation in Spain and the Spanish Ministry for Industry, Commerce and Tourism through Avanza Program. This work was partially supported by Galician Innovation Agency (GAIN) through Conecta-PEME Program-3rd edition (GERIA-TIC Project: IN852A 2016/10); the Carlos III Health Institute from the Spanish National Plan for Scientific and Technical Research and Innovation 2013–2016 (CICLOGEN: PI17/01826) and the European Regional Development Funds (FEDER)—"A way to build Europe". In addition, this work has received financial support from the Xunta de Galicia (Centro singular de investigación de Galicia accreditation 2016–2019) and the European Union (ED431G/01; GRC2014/049), the Galician Network of Drugs R+D REGID (ED431D 2017/16), and the Galician Network for Colorectal Cancer Research-REGICC (ED431D 2017/23), funded by Xunta de Galicia.

Conflicts of Interest: The authors declare no conflict of interest. The founding sponsors had no role in the design of the study; in the collection, analyses, or interpretation of data; in the writing of the manuscript, and in the decision to publish the results

References

1. Bölte, S. Computer-based intervention in autism spectrum disorders. In *Focus on Autism Research*; Ryaskin, O.T., Ed.; Nova Biomedical: New York, NY, USA, 2004; pp. 247–260.
2. Barry, M.; Pitt, I. Interaction design: A multidimensional approach for learners with autism. In Proceedings of the 5th International Conference for Interacting Design and Children [Internet], Tampere, Finland, 7–9 June 2006; pp. 33–36. Available online: http://dl.acm.org/citation.cfm?id=1139086 (accessed on 7 September 2018).
3. Bogdashina, O. *Percepción Sensorial en Autismo y Asperger*; Autismo Ávila: Ávila, Spain, 2007.
4. Moore, M.; Calvert, S. Brief report: vocabulary acquisition for children with autism: Teacher or computer instruction. *J. Autism Dev. Disord.* **2000**, *30*, 359–362.
5. Gaffrey, M.S.; Kleinhans, N.M.; Haist, F.; Akshoomoff, N.; Campbell, A.; Courchesne, E.; Müller, R.A. Atypical [corrected] participation of visual cortex during word processing in autism: An fMRI study of semantic decision. *Neuropsychologia* **2007**, *45*, 1672–1684.
6. Grandin, T. *Pensar con Imágenes: Mi Vida Con el Autismo*; Alba: Barcelona, Spain, 2006.
7. Kana, R.K.; Keller, T.A.; Cherkassky, V.L.; Minshew, N.J.; Just, M.A. Sentence comprehension in autism: Thinking in pictures with decreased functional connectivity. *Brain* **2006**, *129*, 2484–2493.
8. Mesibov, G.; Howley, M. *El Acceso al Currículo Para Alumnos Con TEA*; TEACCH Program; Autismo Ávila: Ávila, Spain, 2010.
9. Stromer, R.; Kimball, J.W.; Kinney, E.M.; Taylor, B.A. Activity schedules, computer technology, and teaching children with autism spectrum disorders. *Focus Autism Other Dev. Disabl.* **2006**, *21*, 14–24.
10. Tomchek, S.D.; Dunn, W. Sensory processing in children with and without autism: a comparative study using the short sensory profile. *Am. J. Occup. Ther.* **2007**, *61*, 190–200.

proceedings

MDPI

Extended Abstract

When Diversity Met Accuracy: A Story of Recommender Systems [†]

Alfonso Landin * [ID]**, Eva Suárez-García** [ID] **and Daniel Valcarce** [ID]

Department of Computer Science, University of A Coruña, 15071 A Coruña, Spain;
eva.suarez.garcia@udc.es (E.S.-G.); daniel.valcarce@udc.es (D.V.)
*   Correspondence: alfonso.landin@udc.es; Tel.: +34-881-01-1276
†   Presented at the XoveTIC Congress, A Coruña, Spain, 27–28 September 2018.

Published: 14 September 2018

check for updates

Abstract: Diversity and accuracy are frequently considered as two irreconcilable goals in the field of Recommender Systems. In this paper, we study different approaches to recommendation, based on collaborative filtering, which intend to improve both sides of this trade-off. We performed a battery of experiments measuring precision, diversity and novelty on different algorithms. We show that some of these approaches are able to improve the results in all the metrics with respect to classical collaborative filtering algorithms, proving to be both more accurate and more diverse. Moreover, we show how some of these techniques can be tuned easily to favour one side of this trade-off over the other, based on user desires or business objectives, by simply adjusting some of their parameters.

Keywords: recommender systems; collaborative filtering; diversity; novelty

---

1. Introduction

Over the years the user experience with different services has shifted from a proactive approach, where the user actively look for content, to one where the user is more passive and content is suggested to her by the service. This has been possible due to the advance in the field of recommender systems (RS), making it possible to make better suggestions to the users, personalized to their preferences.

Most of the research on the field focuses on the accuracy as the main objective of the systems. For example, the Netflix Prize goal was to improve the accuracy of Cinematch (Netflix recommendation system) by 10%, measured by the root mean squared error of the predictions. This competition fuelled the research and several advances came from it. However, in the wake of the results, studies have proven the inadequacy of this measure when it comes to the top-n recommendation task [1], introducing the use of IR metrics, such as precision or the normalized discounted cumulative gain (nDCG), to assess the performance of the system. To introduce these measures non-rated items are considered as non relevant. It has been acknowledged that making this consideration may underestimate the true metric value; however, it provides a better estimation of the recommender quality [2].

Other studies have also pointed out the convenience of measuring different properties of recommender systems such as diversity or novelty [3,4]. A system that is able to produce novel recommendations increases the probability of suggesting items to a user that would not have discovered by herself; this property is called serendipity. This quality is often associated with user satisfaction [5], but it is difficult to measure, usually involving online experiments. We use novelty as a proxy to measure this property. Being able to produce diverse recommendations, that make use of the full catalogue of items instead of focusing on the more popular ones, is usually an added benefit to a recommender system. Diversity is highly appreciated by vendors [6,7].

We analysed the performance of a couple of memory-based recommender systems, both using four different clustering techniques to compute the neighbourhoods. This performance was evaluated in term of precision, diversity and novelty metrics. We also analysed how the systems perform with different values of their parameters, with the intent of showing how the performance of the systems with respect to the trade-off between accuracy and diversity/novelty can be tuned to suit the needs of the user or the business objectives.

2. Materials and Methods

We conducted a series of experiments in order to analyse the trade-off between accuracy, diversity and novelty in Recommender Systems.

2.1. Algorithms

We choose two memory-based based algorithms to analyse their performance. The first one, Weighted Sum Recommender (WSR), is a formulation of the classic user based recommender that stands out for its simplicity and performance [8]. The second one is an adaptation of Relevance-based Language Model (frequently abbreviated as Relevance Models or RM), used in text retrieval to perform pseudo relevance feedback [9]. In particular, we used the RM2 approach, which showed superior performance than RM1 [10].

Both algorithms use the notion of the neighbourhood of a user to perform their calculations. Intuitively, they decide to recommend or not an item based on the preferences of other users that are considered similar to the active one. We explored four clustering techniques to calculate these neighbourhoods with both algorithms. The first one, k-Nearest Neighbours (k-NN), is a well-known technique commonly used with neighbourhood based algorithms [11]. As a second method, we also tested a modification of the k-NN technique, inverted nearest neighbours (k-iNN), that claim to improve both novelty and accuracy [12]. Another technique we used was Posterior Probabilistic Clustering [13], in particular the model that uses the K-L divergence cost function (PPC2). Lastly, we used the Normalized Cuts (NC), a technique used in image segmentation [14], adapted to partition users into clusters. These last two techniques are hard clustering techniques, where a user can only be part of a single cluster. On the contrary, the first two are soft clustering techniques, meaning that a user can be in more than one cluster at the same time. These two methods also make use of a similarity measure, that has to be defined independently. For our research, we used the cosine similarity in both cases.

2.2. Evaluation Protocol

We report out result only on the MovieLens 100k dataset, given the space constraints, although similar trends have been observed in other collections. This is a very popular public dataset for evaluating collaborative filtering methods. It contains 100,000 ratings that 943 users gave to 1682 items. We used the splits provided by the collection to perform 5-fold cross-evaluation.

To evaluate de effectiveness of the recommendations we used the Normalized Discounted Cumulative Gain (nDCG), using the standard formulation as described in [15] with ratings as graded relevance judgements. In our experiments, only items with a rating of 4.0 or higher are considered relevant when evaluating. To assess the diversity of the recommendations we use the inverse of the Gini index [6]. When a value of the index is 0 it signifies that a single item is being recommended to all users. A value of 1 means that all items are recommended equally to all the users. To evaluate the novelty we use the mean self-information (MSI) [16]. All the metrics are evaluated at a cut-off of 10. We do this because we are interested in evaluating the quality of the top recommendations.

3. Results

We tested all the combinations of recommender and clustering techniques. For the soft clustering methods (k-NN and k-iNN) we varied the number of neighbours between 25 and 200. For the hard clustering techniques (PPC2 and NC) we obtained the results modifying the number of clusters between 10 and 100. The results in terms of accuracy (nDCG), diversity (Gini) and novelty (MSI) can be observed in Figure 1.

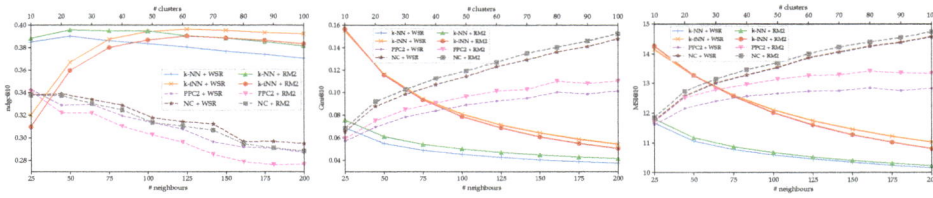

Figure 1. Values of nDCG@10, Gini@10 and MSI@10 of all studied algorithms when varying the number of clusters or neighbours.

When it comes to accuracy alone both *k*-NN and *k*-iNN show a superior performance when compared to the hard clustering methods, offering both similar results in term of nDCG. For these the type of recommender that offers the best results varies. *k*-NN obtains better results with the RM2 algorithm. In the case of *k*-iNN, it is the WSR algorithm that gets the better results.

In the case of the diversity and novelty results, it can be observed that most of the time tuning a method to provide more accurate results leads to a decrease in these other to measures. This is not always true, as can be seen with the soft clustering techniques, when increasing the numbers of neighbours too much leads to decreases in accuracy, diversity and novelty. It can also be seen that different algorithms can obtain different levels of diversity and novelty at the same level of accuracy. In this regard, the *k*-iNN method shows superior levels of diversity and novelty when compared to the *k*-NN technique at similar levels of accuracy, confirming the claim of their proponents.

4. Discussion

Results show that the intuition that during the process of tuning a recommender raising the accuracy leads to decreases in novelty and diversity holds most of the time, but there can be situations when this is no longer true, and the performance of the system moves in the same direction for all the metrics when changing a parameter.

But the results also show that the choice of algorithms is important when it comes to improving the properties of the system. It is possible to improve the performance of the system in diversity and novelty, while maintaining similar levels of accuracy. It is also possible to tune the system to balance how well it performs in all the metrics. This is a multi-objective problem and a trade off must be chosen, either by a priori setting the weight that each measure has, or by choosing any of the possible combination of parameters from the values in the Pareto front.

Funding: This work has received financial support from project TIN2015-64282-R (MINECO/ERDF), project GPC ED431B 2016/035 (Xunta de Galicia) and accreditation ED431G/01 (Xunta de Galicia/ERDF). The first author also acknowledges the support of grant FPU17/03210 (MECD). The third author also acknowledges the support of grant FPU014/01724 (MECD).

Conflicts of Interest: The authors declare no conflict of interest. The founding sponsors had no role in the design of the study; in the collection, analyses, or interpretation of data; in the writing of the manuscript, and in the decision to publish the results.

References

1. Cremonesi, P.; Koren, Y.; Turrin, R. Performance of Recommender Algorithms on Top-N Recommendation Tasks. In Proceedings of the 4th ACM Conference on Recommender Systems (RecSys'10), Barcelona, Spain, 26–30 September 2010; ACM: New York, NY, USA, 2010; pp. 39–46, doi:10.1145/1864708.1864721.

2. McLaughlin, M.R.; Herlocker, J.L. A collaborative filtering algorithm and evaluation metric that accurately model the user experience. In Proceedings of the 27th Annual International Conference on Research and Development in Information Retrieval (SIGIR'04), Sheffield, UK, 25–29 July 2004; ACM Press: New York, NY, USA, 2004; p. 329, doi:10.1145/1008992.1009050.

3. Herlocker, J.L.; Konstan, J.A.; Terveen, L.G.; Riedl, J.T. Evaluating collaborative filtering recommender systems. *ACM Trans. Inf. Syst.* **2004**, *22*, 5–53, doi:10.1145/963770.963772.

4. McNee, S.M.; Riedl, J.; Konstan, J.A. Being Accurate is Not Enough: How Accuracy Metrics have hurt Recommender Systems. In Proceedings of the CHI'06 Extended Abstracts on Human Factors in Computing Systems (CHI EA'06), Montréal, QC, Canada, 22–27 April 2006; ACM Press: New York, NY, USA, 2006; p. 1097, doi:10.1145/1125451.1125659.

5. Ge, M.; Delgado-Battenfeld, C.; Jannach, D. Beyond Accuracy: Evaluating Recommender Systems by Coverage and Serendipity. In Proceedings of the Fourth ACM Conference on Recommender Systems (RecSys'10), Barcelona, Spain, 26–30 September 2010; pp. 257–260, doi:10.1145/1864708.1864761.

6. Fleder, D.; Hosanagar, K. Blockbuster Culture's Next Rise or Fall: The Impact of Recommender Systems on Sales Diversity. *Manag. Sci.* **2009**, *55*, 697–712, doi:10.1287/mnsc.1080.0974.

7. Valcarce, D.; Parapar, J.; Álvaro Barreiro. Item-based relevance modelling of recommendations for getting rid of long tail products. *Knowl.-Based Syst.* **2016**, *103*, 41–51, doi:10.1016/j.knosys.2016.03.021.

8. Valcarce, D.; Parapar, J.; Barreiro, A. Efficient Pseudo-Relevance Feedback Methods for Collaborative Filtering Recommendation. In Proceedings of the European Conference on Information Retrieval (ECIR'16), Padua, Italy, 20–23 March 2016; pp. 602–613, doi:10.1007/978-3-319-30671-1_44.

9. Lavrenko, V.; Croft, W.B. Relevance based language models. In Proceedings of the 24th Annual International ACM SIGIR Conference on Research and Development in Information Retrieval (SIGIR'01), New Orleans, LA, USA, 9–12 September 2001; ACM Press: New York, NY, USA, 2001; pp. 120–127, doi:10.1145/383952.383972.

10. Parapar, J.; Bellogín, A.; Castells, P.; Barreiro, A. Relevance-based language modelling for recommender systems. *Inf. Process. Manag.* **2013**, *49*, 966–980, doi:10.1016/j.ipm.2013.03.001.

11. Ning, X.; Desrosiers, C.; Karypis, G. A Comprehensive Survey of Neighborhood-Based Recommendation Methods. In *Recommender Systems Handbook*, 2nd ed.; Ricci, F., Rokach, L., Shapira, B., Eds.; Springer: Boston, MA, USA, 2015; pp. 37–76, doi:10.1007/978-1-4899-7637-6_2.

12. Vargas, S.; Castells, P. Improving Sales Diversity by Recommending Users to Items. In Proceedings of the 8th ACM Conference on Recommender Systems (RecSys'14), Foster City, CA, USA, 6–10 October 2014; ACM: New York, NY, USA, 2014; pp. 145–152, doi:10.1145/2645710.2645744.

13. Zhang, Z.Y.; Li, T.; Ding, C.; Tang, J. An NMF-framework for Unifying Posterior Probabilistic Clustering and Probabilistic Latent Semantic Indexing. *Commun. Stat.-Theory Methods* **2014**, *43*, 4011–4024, doi:0.1080/03610926.2012.714034.

14. Shi, J.; Malik, J. Normalized cuts and image segmentation. *IEEE Trans. Pattern Anal.* **2000**, *22*, 888–905, doi:10.1109/34.868688.

15. Wang, Y.; Wang, L.; Li, Y.; He, D.; Chen, W.; Liu, T.Y. A Theoretical Analysis of NDCG Ranking Measures. In Proceedings of the 26th Annual Conference on Learning Theory (COLT'13), Princeton, NJ, USA, 12–14 June 2013; pp. 1–30.

16. Zhou, T.; Kuscsik, Z.; Liu, J.G.; Medo, M.; Wakeling, J.R.; Zhang, Y.C. Solving the apparent diversity-accuracy dilemma of recommender systems. *Proc. Natl. Acad. Sci. USA* **2010**, *107*, 4511–4515, doi:10.1073/pnas.1000488107.

proceedings

MDPI

Extended Abstract

Brain-Computer Interfaces for Internet of Things †

Francisco Laport *, Francisco J. Vazquez-Araujo, Paula M. Castro and Adriana Dapena

Department of Computer Engineering, Universidade da Coruña, 15071 A Coruña, Spain;
fjvazquez@udc.es (F.J.V.-A.); paula.castro@udc.es (P.M.C.); adriana.dapena@udc.es (A.D.)
* Correspondence: francisco.laport@udc.es; Tel.: +34-981167000
† Presented at the XoveTIC Congress, A Coruña, Spain, 27–28 September 2018.

Published: 17 September 2018

Abstract: A brain-computer interface for controlling elements commonly used at home is presented in this paper. It includes the electroencephalography device needed to acquire signals associated to the brain activity, the algorithms for artefact reduction and event classification, and the communication protocol.

Keywords: Brain-Computer Interfaces; Internet of Things; Smart Home

1. Introduction

Brain-Computer Interfaces (BCI) are defined as the communication systems that monitor cerebral activity and translate certain characteristics, corresponding to user intentions, to commands for device control. Current research is focused on the potential of Electroencephalography (EEG) [1] devices to capture the brain activity associated to the user intentionality. The acquired signals are then translated to external components [2]. The integration of BCI and Internet of Things (IoT) for Smart Home (SH) is a promising and emerging technique to make home environments comfortable and accessible, automating and optimizing the use of appliances, like TV sets, air conditioners, light bulbs, etc. [3].

In this paper, we include two relevant parts of an EEG system for IoT. Firstly, we introduce some details of a self-designed EEG interface with single channel and two methods for the analysis and classification of the obtained data. Secondly, we explain the integration of this device with an IoT system based on the widely used protocol termed as Message Queue Telemetry Transport (MQTT) [4].

2. Developed System

For the integration of BCI and SH, we have developed the system shown in Figure 1. The first element of the proposed system is the device for EEG data acquisition. This captures the user's brain activity using a single channel, whose location on the scalp depends on the intentionality to be captured. EEG data measurements are obtained with a sampling frequency of 128 Hz in the frequency range from 4.7 to 22 Hz. The raw data captured by the EEG device are then analyzed by an ESP32 microcontroller module [5] to determine the user state. As an example of application of our system, we will consider classification in open/closed eyes. It is well known that EEG alpha activity (8–13 Hz) increases for normal individuals during a closed eyes situation and is suppressed with visual stimulation. According to these studies, the signal processing unit will be able to determine the user eye state considering the power of alpha (α) and beta (β). The closed eyes state is associated to a β/α ratio lower than a certain threshold level due to higher alpha power. Conversely, the open eyes state corresponds to a β/α ratio higher than that threshold level due to the smaller alpha power.

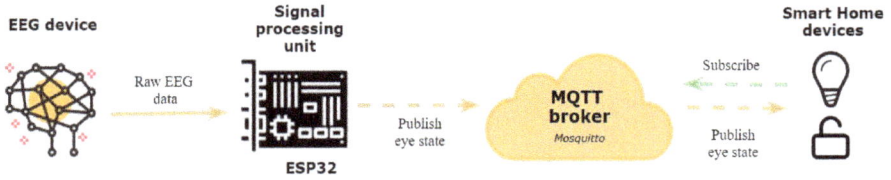

Figure 1. Scheme of the system developed for the integration of BCI and IoT.

Once the eyes state is determined by the system, it must be communicated to the different devices of the SH. For this purpose, the MQTT protocol is used. It is a widely used communication protocol in IoT applications which applies the publish/subscribe pattern. This protocol traduces the user intentionality to control commands. For instance, the state of closed eyes during a "a priori" fixed time interval could correspond to modify the environment (home or hospital room, for example) to a "sleeping" mode (turn off light and TV, etc.).

3. Experiments

EEG data of four different individuals have been recorded and analyzed. The EEG was performed using only one channel placed at the FP2 position according to the 10/20 system [6]. Eighteen trials of 20 s for each eye state were obtained from the participants. We propose two methods for determining the eye state: method 1, which obtains the alpha and beta bands using band-pass filters and then computes the β/α ratio; method 2, which computes the Fast Fourier Transform (FFT) of the acquired signal for the frequencies corresponding to alpha and beta bands and then computes the β/α ratio. Figure 2 shows the values for closed and open eyes averaging all these trials from the training period and also the threshold levels obtained for everyone. As you can see in the subfigures corresponding to both methods, there are significant differences between participants.

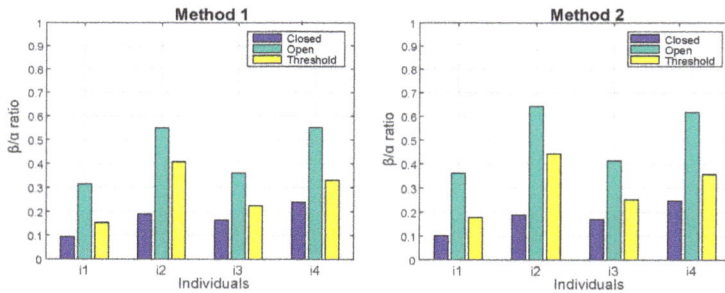

Figure 2. β/α ratio of each eye state for methods 1 and 2 and its corresponding threshold level.

In the test period, we have used the thresholds previously calculated during the training working for classification of 24 trials per participant obtained from eight measurements per day along three days. Table 1 shows the accuracy obtained using both methods. As you can see in the table, the good performance of both methods is evident from percentages higher than 95% for all users.

Table 1. Accuracy of the proposed methods for test working.

METHOD	SUBJECT 1	SUBJECT 2	SUBJECT 3	SUBJECT 4
METHOD 1	100%	100%	95%	100%
METHOD 2	100%	100%	95%	100%

4. Conclusions

The proposed system for the integration of BCI and IoT is formed by a single channel EEG device, a signal processing module to determine the user's eye state and the MQTT protocol for the distribution of the extracted knowledge among the connected devices. The results show that the proposed system achieves high accuracy.

Author Contributions: F.L. has implemented all software used in this paper, performed the experiments, analyzed the data and wrote the paper; F.J.V.-A. has developed the EEG hardware; P.M.C. and A.D. have designed the experiments and head the research.

Acknowledgments: This work has been funded by a contract granted by Xunta de Galicia (Francisco Laport).

Conflicts of Interest: The authors declare no conflict of interest.

References

1. Sanei, S.; Chambers, J. A. *EEG Signal Processing*; John Wiley & Sons: Chichester, UK, 2013.
2. Vidal, J.J. Real-time detection of brain events in EEG. *Proc. IEEE* **1977**, *65*, 633–641, doi:10.1109/PROC.1977.10542.
3. Mathe, E.; Spyrou, E. Connecting a Consumer Brain-Computer Interface to an Internet-of-Things Ecosystem. In Proceedings of the ACM PETRA'16, Corfu Island, Greece, 1–9 June 2016, doi:10.1145/2910674.2935844.
4. MQTT. Available online: http://mqtt.org/ (accessed on 14 July 2018).
5. ESP32-WROOM-32 Datasheet. Available online: https://www.espressif.com/sites/default/files/documentation/esp32-wroom-32_datasheet_en.pdf (accessed on 14 July 2018).
6. Jasper, H.H. The ten-twenty electrode system of the International Federation. *Electroencephalogr. Clin. Neurophysiol.* **1958**, *10*, 370–375.

![proceedings logo] *proceedings*

MDPI

Extended Abstract

Fluid Region Analysis and Identification via Optical Coherence Tomography Image Samples [†]

Plácido L. Vidal [1,2,]*, Joaquim de Moura [1,2], Jorge Novo [1,2], José Rouco [1,2] and Marcos Ortega [1,2]

[1] Department of Computer Science, University of A Coruña, 15071 A Coruña, Spain;
joaquim.demoura@udc.es (J.d.M.); jnovo@udc.es (J.N.); jrouco@udc.es (J.R.); mortega@udc.es (M.O.)

[2] CITIC-Research Center of Information and Communication Technologies, University of A Coruña, 15071 A Coruña, Spain

* Correspondence: placido.francisco.lizancos.vidal@udc.es; Tel.: +34-881-01-13-30

[†] Presented at the XoveTIC Congress, A Coruña, Spain, 27–28 September 2018.

Published: 17 September 2018

Abstract: The work herein proposed presents a methodology which aims to identify cystoid regions using OCT scans. This method obtained satisfactory results detecting cystoid regions with different levels of complexity without needing any preprocessing nor candidate filtering steps.

Keywords: computer-aided diagnosis; retinal imaging; optical coherence tomography; intraretinal cystoid region characterization; feature selection; classification

1. Introduction

Optical Coherence Tomography (OCT) is a non-invasive medical imaging modality that provides morphological information about the retinal tissues. This information is commonly used in the early diagnosis and analysis of patients with potential eye and systemic diseases as, for reference, Diabetic Retinopathy (DR), Diabetic Macular Edema (DME) and Age-related Macular Degeneration (AMD), three of the leading causes of blindness in adults of working age in developed countries. Given the global relevance of this topic, an accurate identification of any present cystoid region is crucial to perform an adequate diagnosis, treatment, prevention and rehabilitation.

2. Methodology

In this work, we propose a novel methodology for the automatic identification and characterization of the intraretinal cystoid fluid regions using OCT images [1]. To achieve this, we analyzed a complete and heterogeneous set of 326 intensity and texture descriptors. The most relevant features were selected using the Relief-F and L0 feature selectors and tested with the Linear Discriminant Classifier, the Quadratic Discriminant Classifier and k Nearest Neighbors with k = 5.

3. Results

The proposed methodology was tested using 51 OCT images obtained with a CIRRUS© HD-OCT confocal laser ophthalmoscope (Carl Zeiss Meditec, Inc., Dublin, California). From these images, a total of 723 samples from cystoid and non-cystoid regions were extracted, using a sample size of 51 × 51. This sample size was empirically determined to be large enough to detect both big and small cystoid bodies in the retinal layers.

The methodology correctly identified the intraretinal cystoid fluid regions with a satisfactory accuracy of 90.6%. As shown in Figure 1, both fluid regions and non-cystoid regions are detected by the system despite the multiple complications.

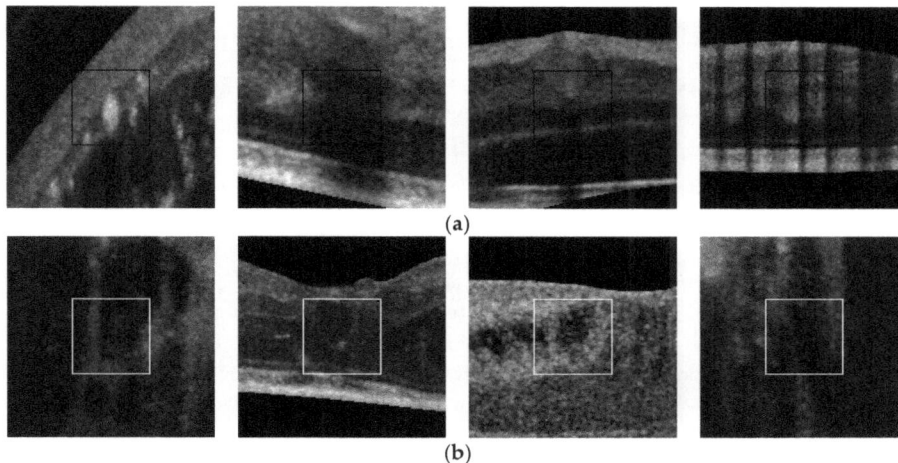

Figure 1. Results obtained with different levels of complexity. (**a**) Non-cystoid regions; (**b**) Cystoid regions.

Author Contributions: All authors contributed equally.

Acknowledgments: This work is supported by the Instituto de Salud Carlos, III, Government of Spain and FEDER funds of the European Union through the PI14/02161 and the DTS15/00153 research projects and by the Ministerio de Economía y Competitividad, Government of Spain through the DPI2015-69948-R research project.

Conflicts of Interest: The authors declare no conflict of interest. The founding sponsors had no role in the design of the study; in the collection, analyses, or interpretation of data; in the writing of the manuscript, and in the decision to publish the results.

Reference

1. De Moura, J.; Vidal, P.L.; Novo, J.; Rouco, J.; Ortega, M. Feature definition, analysis and selection for cystoid region characterization in optical coherence tomography. *Procedia Comput. Sci.* **2017**, *112*, 1369–1377, doi:10.1016/j.procs.2017.08.043.

proceedings

MDPI

Extended Abstract

Nonparametric Inference in Mixture Cure Models [†]

Ana López-Cheda [1,*] [iD], Ricardo Cao [1], Mª Amalia Jácome [1] and Ingrid Van Keilegom [2]

[1] Department of Mathematics, University of A Coruña, 15071 A Coruña, Spain; rcao@udc.es (R.C.); majacome@udc.es (M.A.J.)

[2] ORSTAT, KU Leuven, 3000 Leuven, Belgium; ingrid.vankeilegom@kuleuven.be

[*] Correspondence: ana.lopez.cheda@udc.es; Tel.: +34-981-167-000 (ext. 1301)

[†] Presented at the XoveTIC Congress, A Coruña, Spain, 27–28 September 2018.

Published: 17 September 2018

Abstract: A completely nonparametric method for the estimation of mixture cure models is proposed. Nonparametric estimators for the cure probability (incidence) and for the survival function of the uncured population (latency) are introduced. In addition, a bootstrap bandwidth selection method for each nonparametric estimator is considered. The methodology is applied to a dataset of colorectal cancer patients from the University Hospital of A Coruña (CHUAC). Furthermore, a nonparametric covariate significance test for the incidence is proposed. The test is extended to non-continuous covariates: binary, discrete and qualitative, and also to contexts with a large number of covariates. The method is applied to a sarcomas dataset from the University Hospital of Santiago (CHUS).

Keywords: bandwidth selection; bootstrap; censored data; kernel estimation; survival analysis

---

1. Introduction

In the last two decades there has been a remarkable progress in cancer treatments, which led to longer patient survival and improved their quality of life. Consequently, a spate of statistical research to develop cure models arose. These models are useful tools to analyze and describe survival data with long-term survivors, since they express and predict the prognosis of a patient considering, as a novelty, the real possibility that the subject may never experience the event of interest. Cure models allow to estimate the cured proportion, $1 - p(x)$, and also the probability of survival of the uncured patients up to a given time point, or latency, $S_0(t|x)$. In the literature, ref. [1] proposed the nonparametric incidence estimator: $1 - \hat{p}_h(x) = \hat{S}_h(T^1_{\max}|x)$, where $\hat{S}_h()$ is the conditional Kaplan-Meier estimator with bandwidth h, and $T^1_{\max}$ is the largest uncensored failure time. The first completely nonparametric approach in mixture cure models was proposed by [2], who introduced the nonparametric latency estimator: $\hat{S}_{0,b}(t|x) = \frac{\hat{S}_b(t|x) - (1 - \hat{p}_b(x))}{\hat{p}_b(x)}$, studied in detail by [3]. Furthermore, in cancer studies it is interesting to test if a covariate has some influence on the cure rate or on the survival time of the susceptible patients. Since no significance testing has been proposed yet for nonparametric cure models, this important gap is filled with the proposal of a covariate significance test for the incidence. This test allows to identify which covariates must be included in the incidence in a mixture cure model. Following [4], the proposed statistics is based on the process:

$$T_n(z) = \frac{1}{n} \sum_{i=1}^{n} \left(\hat{\eta}_i - \left(\frac{1}{n} \sum_{j=1}^{n} \hat{\eta}_j \right) \right) I(Z_i \leq z),$$

where n is the sample size, $\hat{\eta}_i$ is an estimator of the cure indicator for each individual, and Z is the covariate. Possible test statistics are the Cramér-von Mises (CvM) or the Kolmogorov-Smirnov (KS) tests. Moreover, the test statistic null distribution is approximated by bootstrap, using an independent naive resampling. For the case with an m-dimensional covariate, **Z**, the method consists of considering

m hypotheses in H_0 to be tested independently. In order to control the false discovery rate, the approach by [5] to problems of multiple significance testing is studied. In addition, to achieve the family wise error rate control, the conservative method by [6] is considered.

Application to Medical Data

The proposed methodology is applied to a dataset of 414 colorectal cancer patients from CHUAC. The goal is to estimate the cure rate as a function of the stage (from 1 to 4) and the age. The event of interest is the death due to colorectal cancer, and the censoring percentage is between 30.77% (Stage 4) and 70.97% (Stage 1). Figure S1 in the Supplementary Materials shows that the effect of the age on the cure rate changes with the stage. For example, in Stage 1, patients have a probability of survival between 0.25 and 0.65, depending on the age; whereas in Stage 3, for patients above 60, in a 10 years gap that probability decreases considerably from 0.4 to almost 0. The latency estimation for three specific ages is shown in Figure S2 in the Supplementary Materials. For Stages 1–2, the age does not seem to be determining for the survival of the uncured patients. On the contrary, for Stages 3–4, the latency estimation varies considerably depending on the age. For example, the probability that the follow-up time since the diagnostic until death is larger than 4.5 years is around 0.2 for patients with ages 35 and 50, whereas for 80 year old patients, that probability is larger than 0.4.

Moreover, a dataset related to patients with sarcomas, provided by CHUS, is studied. It consists of 261 observations with 372,420 covariates with information about DNA methylations and 32 covariates with clinical data. The event of interest is the death due to sarcomas, and a total of 195 observations are censored. Regarding the conservative method, the results show that only one covariate is significant for the cure rate: "Year of initial pathologic diagnosis". With respect to the non-conservative alternative, the results for $B = 10^5$ bootstrap resamples show that for the CvM statistic, there are 14,182 significant covariates and 650 non-conclusive covariates, which need to be considered again in the next iteration of the process. For the KS statistic, there are 12,411 significant covariates, and 608 non-conclusive covariates. The program is still running for $B = 10^6$ bootstrap resamples.

2. Discussion

Mixture cure models have been usually estimated using parametric or semiparametric methods. A completely nonparametric approach for the estimation in mixture cure models is introduced, and a nonparametric covariate significance test for the probability of cure in mixture cure models is proposed. The methodology, that can be applied to any type of covariates and to high dimensional datasets, is illustrated with medical data. Specifically, the nonparametric incidence and latency estimators are applied to a dataset related to colorectal cancer patients from CHUAC. The incidence in Stages 1 and 2 is higher than in Stages 3 and 4 due to the fact that most of the surgeries in initial stages have healing purposes, whereas in advanced stages, surgeries are usually palliative treatments, and therefore the cure rate is lower. Furthermore, the latency estimation in Stages 3 and 4 is higher for 80 year old patients than for younger patients. The reason is that when a colorectal cancer is diagnosed in a young patient, it is usually in an advanced stage and with worse prognosis, since the cancer cells are more active in young individuals. Regarding the proposed covariate significance test for the incidence with the high dimensional dataset of sarcomas, the results differ for the conservative and the non-conservative approaches.

3. Materials

An R package is being developed with all the techniques proposed, including the implementation of the nonparametric incidence and latency estimators, as well as the covariate significance tests for different types of data: continuous, discrete, binary and qualitative, and for a high dimensional covariate vector. This R package will be uploaded in the Comprehensive R Archive Network (CRAN).

Supplementary Materials: The following are available online at http://www.mdpi.com/2504-3900/2/18/1181/s1. Figure S1: Nonparametric cure rate estimation for the different stages, Figure S2: Nonparametric latency estimation for the different stages.

Author Contributions: Conceptualization, R.C., M.A.J. and I.V.K.; Methodology, A.L.-C., R.C., M.A.J. and I.V.K.; Software, A.L.-C.; Validation, A.L.-C., R.C. and M.A.J.; Investigation, A.L.-C., R.C. and M.A.J.; Writing—Original Draft Preparation, A.L.-C.; Writing—Review & Editing, A.L.-C., R.C. and M.A.J.; Supervision, R.C. and M.A.J.

Funding: The first author was sponsored by the Spanish FPU (Formación de Profesorado Universitario) Grant from MECD (Ministerio de Educación, Cultura y Deporte) with reference FPU13/01371. The work has been partially carried out during two visits at the Université catholique de Louvain. The first stay was financed by INDITEX and the second one was supported by the research group MODES (Modelización, Optimización e Inferencia Estadística). All the authors acknowledge partial support by the MICINN (Ministerio de Ciencia e Innovación) Grant MTM2011-22392 and the MINECO (Ministerio de Economía y Competitividad) Grant MTM2014-52876-R. The first three authors acknowledge partial support of Xunta de Galicia (Grupos de Referencia Competitiva CN2012/130, ED431C-2016-015 and Centro Singular de Investigación de Galicia ED431G/01), all of them through the ERDF (European Regional Development Fund). Financial support from the European Research Council (2016-2021, Horizon 2020 / ERC grant agreement No. 694409) is gratefully acknowledged.

Acknowledgments: The authors are grateful to S. Pértega and S. Pita, at CHUAC, for providing the colorectal cancer dataset; and to A. Díaz-Lagares, at CHUS, for providing the sarcomas dataset.

Conflicts of Interest: The authors declare no conflict of interest. The founding sponsors had no role in the design of the study; in the collection, analyses, or interpretation of data; in the writing of the manuscript, and in the decision to publish the results.

References

1. Xu, J.; Peng, Y. Nonparametric cure rate estimation with covariates. *Can. J. Stat.* **2014**, *42*, 1–17. doi:10.1002/cjs.11197.

2. López-Cheda, A.; Cao, R.; Jácome, M.A.; Van Keilegom, I. Nonparametric incidence estimation and bootstrap bandwidth selection in mixture cure models. *Comput. Stat. Data Anal.* **2017**, *105*, 144–165. doi:10.1016/j.csda.2016.08.002.

3. López-Cheda, A.; Jácome, M.A.; Cao, R. Nonparametric latency estimation for mixture cure models. *Test* **2017**, *26*, 353–376. doi:10.1007/s11749-016-0515-1.

4. Delgado, M.A.; González-Manteiga, W. Significance testing in nonparametric regression based on the bootstrap. *Ann. Stat.* **2001**, *29*, 1469–1507. doi:10.1214/aos/1013203462.

5. Benjamini, Y.; Hochberg, Y. Controlling the false discovery rate: A practical and powerful approach to multiple testing. *J. R. Stat. Soc. Ser. B Stat. Methodol.* **1995**, *57*, 289–300. doi:10.2307/2346101.

6. Benjamini, Y.; Yekutieli, D. The control of the false discovery rate in multiple testing under dependency. *Ann. Stat.* **2001**, *29*, 1165–1188. doi:10.1214/aos/1013699998.

proceedings

MDPI

Extended Abstract

Numerical Simulation of the Dynamics of Listeria Monocytogenes Biofilms †

Eva Balsa-Canto [1], **Alejandro López-Núñez** [2,*] **and Carlos Vázquez** [2]

[1] Process Engineering Group, IIM-CSIC, Spanish Council for Scientific Research, Eduardo Cabello 6, 36208 Vigo, Spain; ebalsa@iim.csic.es
[2] Department of Mathematics, University of A Coruña, Campus Elviña s/n, 15071 A Coruña, Spain; carlosv@udc.es
* Correspondence: alopeznu@gmail.com; Tel.: +34-981-167-000
† Presented at the XoveTIC Congress, A Coruña, Spain, 27–28 September 2018.

Published: 18 September 2018

Abstract: A biofilm is a layer of microorganisms attached to a surface and protected by a matrix of exopolysaccharides. Biofilm structures difficult the removal of microorganisms, thus the study of the type of structures formed throughout a biofilm life cycle is key to design elimination techniques. Also, the study of the inner mechanisms of a biofilm system is of the utmost importance in order to prevent harmful biofilms formation and enhance the properties of beneficial biofilms. This study must be achieved through the combination of mathematical modelling and experimental studies. Our work focuses on the study of biofilms formed by *Listeria monocytogenes*, a pathogen bacteria, specially relevant in food industry. Listeria is highly resistant to biocides and appears in common food surfaces even after decontamination processes. Their biofilms can develop quite different structures, from flat biofilms to clustered or honeycomb structures. In the present work, we develop 1D and 2D models that simulate the dynamics of biofilms formed by different strains of *L. monocytogenes*. All this models are solved with efficient numerical methods and robust numerical techniques, such as the Level Set method. The numerical re sults are compared with the experimental measurements obtained in the Instituto de Investigaciones Marinas, CSIC (Vigo, Spain), and the Micalis Institute, INRA (Massy, France).

Keywords: applied mathematics; numerical simulation; microbiology

---

1. Introduction

Listeria monocytogenes is a pathogenic bacteria responsible for outbreaks of listeriosis. The main mode of transmission of this pathogen to humans is the consumption of contaminated food through contact with unhygienic work surfaces and facilities where *L. monocytogenes* can form biofilms [1].

Biofilms structure determines the main physiological processes related to persistence and resistance. Therefore, structure characterization is critical to design cost effective and environmentally friendly disinfection techniques [2]. Confocal laser scanner microscopy (CLSM) allows for *in vivo* and *in situ* biofilms observation.

In parallel to the experimental studies, the use of efficient mathematical models allows the prediction of the biofilm evolution for particular values of the involved parameters associated to different conditions. Having in view the experimental dynamics of the particular biofilm formed by the L1A1 *L. monocytogenes* strain, we start by considering the most successful 1D continuum model studied in the recent work [2]. With the knowledge acquired in the 1D model, a 2D continuum multi-species model is developed [3] so that we are able to describe several dynamics shown by different *L. monocytogenes* strains. Both models are solved numerically by applying efficient numerical techniques such as Crank-Nicolson schemes, WENO methods or the Level Set method [3,4]. The numerical results that arise are compared with the experimental measurements obtained in the IIM-CSIC (Vigo, Spain), and the Micalis Institute, INRA (Massy, France).

2. One-Dimensional Model

To elucidate the mechanisms explaining the life cycle of the biofilms formed by L1A1 strain we analysed several models until reaching the most successful one [2]. Unknown parameters from the model were estimated using data fitting techniques within the AMIGO2 toolbox [5]. The model is a 1D deterministic reaction-diffusion model. It consist of a set of (non-linear) partial differential equations (PDEs) which describe the spatio-temporal dynamics of biomass and nutrients. The key features of the model are:

- There is a sharp front of biomass at the bulk/solid transition.
- Biomass density can not exceed a maximum bound which is a parameter of the model.
- Biomass production is due to nutrient consumption.
- Nutrients diffuse in the bulk and in the biofilm with different diffusion constants.
- The detachment is related to biofilm ageing.

All in all, the model is described by the following equations:

$$\frac{\partial S}{\partial t} = \frac{\partial}{\partial x}\left(d_1(C)\frac{\partial S}{\partial x}\right) - K_1 SC \tag{1}$$

$$\frac{\partial C}{\partial t} = d_2 \frac{\partial^2 C}{\partial x} + K_3 SC - K_4 \frac{1}{1 + exp(k_d[D_{min} - CBD(t)])} \qquad \text{(Error! No sequence specified.)}$$

completed with appropriate initial and boundary conditions. Equation (1) describes the nutrients dynamics whereas Equation (2) describes the biomass dynamics.

3. Two-Dimensional Model

With the insights provided by the 1D case, a two-dimensional model is built so that it is able to describe the dynamics of the L1A1 strain as well as the clustered or honeycomb patterns presented by other strains such as the CECT 5873 [6]. The proposed model is a deterministic multi-species model of the W-G type. The key hypotheses are:

- Biofilm described as a viscous fluid.
- Nutrients and biomasses concentrations governed by a mass conservation law.
- Active and inactive biomasses are of the same microbial species and incompressible.
- The time scale for the biomass-related processes is much slower that for the nutrients-related.
- Nutrients are diluted in the media. Biomass exists only inside the biofilm.
- The detachment is related to cells death and the degradation of the extracellular DNA, i.e., to biofilm ageing.

All in all, the model is described by the following equations:

$$-\nabla^2 S = -V_1 \hat{h}_T^2 S \in \Omega_{H_b}^\tau \tag{3}$$

$$-\nabla^2 P = V_1 \hat{P} S - V_1 F_D(V_2) \in \Omega_l^\tau \tag{4}$$

$$\acute{U} = -\nabla P \in \Omega_l^\tau \tag{5}$$

$$\partial_\tau \Phi + F_e \|\nabla \Phi\| = 0 \in \Omega_l^\tau \tag{6}$$

$$\partial_\tau V_1 - \nabla P.\nabla V_1 = V_1 \left[\phi S - \left(\frac{1}{1 + exp\left(k_d\left(D_{min} - mean(V_2)\right)\right)} + \varepsilon_2 \right) \right.$$
$$\left. - V_1 \left(\phi S - \frac{1}{1 + exp\left(k_d\left(D_{min} - mean(V_2)\right)\right)} \right) \right] \in \Omega_l^\tau \tag{7}$$

$$\partial_\tau V_2 - \nabla P.\nabla V_2 = V_1 \varepsilon_2 - V_2 \left(V_1 \phi S - \frac{V_1}{1 + exp\left(k_d\left(D_{min} - mean(V_2)\right)\right)} \right) \in \Omega_l^\tau \tag{8}$$

plus appropriate initial and boundary conditions. Equation (3) describes the nutrients dynamics and Equation (4) describes the biofilm expansion growth pressure. Equations (5) and (6) are related to the level set method. Equations (7) and (8) are the active and inactive biomass dynamics equations.

3. Results

After solving both models numerically with the appropriate optimal model parameters, the results yielded are presented in Figures 1 and 2. Starting with the 1D case, Figure 1 it can be noted that biomass thickness is slowed down reaching its peak around 100 h. Nutrients are consumed until the nutrients impairing mechanism starts, preventing biomass from consuming all the nutrients in the domain. Also, it can be observed how the massive detachment happens in the final stage. Results reveal that the model is in clear agreement with the experimental data. Therefore concluding that the life cycle of L1A1 *L. monocytogenes* under the tested experimental conditions may be explained by taking into account impaired nutrients uptake and a massive detachment due to biofilm ageing.

As for the 2D model, Figure 2 shows the different dynamics achieved by the model with appropriate modifications for the model parameters. The results for the flat biofilms correspond to the dynamics of L1A1 *L. monocytogenes* and are in good agreement with the experimental measurements. On the other hand, the results for the clustered biofilms and honeycomb patterns represent the dynamics of the CECT 5873 *L. monocytogenes* strain at different stages of its life cycle, as can be seen in Figure 2, in cases B2 and B4 respectively.

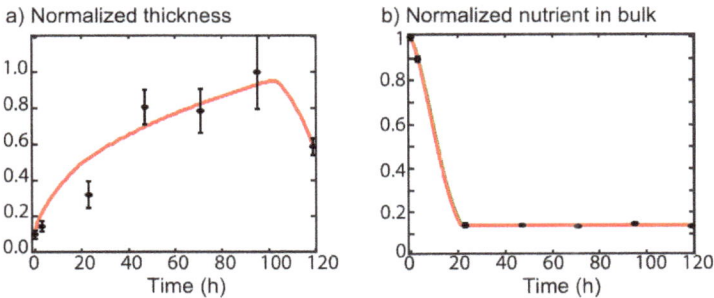

Figure 1. Best fit for the real data of the averaged nutrients and biofilm thickness dynamics predicted by the 1D model. Red line corresponds to the numerical results whereas the black points correspond to the experimental measurements.

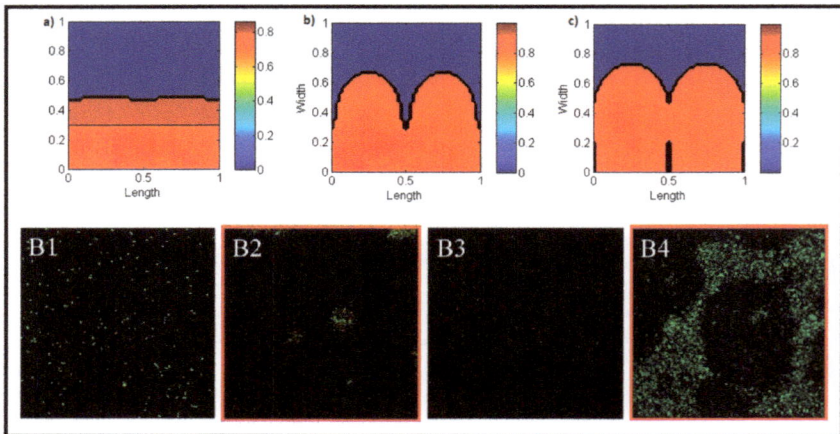

Figure 2. Numerical results for the active biomass in the 2D case. Above: (**a**) flat biofilms; (**b**) clusterd biofilms, (**c**) Honeycomb pattern. Below: CECT 5873 dynamics with IMARIS. Starting with an initial attachment (B1), the biofilm develops a clustered structure (B2). After the apparition of an important quantity of inert biomass (B3), the biofilm develops a honeycomb structure. Source: [6].

Author Contributions: Conceptualization E.B.-C. and C.V.; Methodology E.B.-C., A.L.-N. and C.V.; Software A.L.-N.; Validation E.B.-C.; Formal analysis A.L.-N. and C.V.; Investigation E.B.-C.; Resources E.B.-C.; Supervision E.B.-C. and C.V.

Acknowledgments: This research was funded by FPU grant (ref. FPU13/02191) from the Spanish government, predoctoral grant (ref. PRE/2013/293) from the Galician Government, mobility grant from Centro Singular CITIC, Spanish projects MTM2016-76497-R and MTM2013-47800-C2-1-P, grant GRC2014/044 from the Galician Government and FEDER funds.

Conflicts of Interest: The authors declare no conflict of interest.

References

1. Swaminathan, B.; Gerner-Smidt, P. The epidemiology of human listeriosis. *Microbes Infect.* **2007**, *9*, 1236–1243, doi:10.1016/j.micinf.2007.05.011.
2. Balsa-Canto, E.; Vilas, C.; López-Núñez, A.; Mosquera-Fernández, M.; Briandet, R.; Cabo, M.L.; Vázquez, C. Modeling reveals the role of aging and glucose uptake impairment in L1A1 Listeria monocytogenes biofilm life cycle. *Front. Microbiol.* **2017**, *8*, 2118, doi:10.3389/fmicb.2017.02118.
3. López-Núñez, A. Contributions to Mathematical Modelling and Numerical Simulation of Biofilms. Ph.D. Thesis, Universidade da Coruña, A Coruña, Spain, 2018.
4. Balsa-Canto, E.; López-Núñez, A.; Vázquez, C. Numerical methods for a non-linear reaction-diffusion system modelling a batch culture of biofilm. *Appl. Math. Model.* **2017**, *41*, 164–179, doi:10.1016/j.apm.2016.08.020.
5. Balsa-Canto, E.; Henriques, D.; Gabor, A.; Banga, J.R. AMIGO2, a toolbox for dynamic modeling optimization and control in systems biology. *Bioinformatics* **2016**, *32*, 3357–3359.
6. Mosquera-Fernández, M.; Sánchez-Vizuete, P.; Briandet, R.; Cabo, M.L.; Balsa-Canto, E. Quantitative image analysis to characterize the dynamics of Listeria monocytogenes biofilms. *Int. J. Food Microbiol.* **2016**, *236*, 130–137, doi:10.1016/j.ijfoodmicro.2016.07.015.

proceedings

MDPI

Extended Abstract

Using Discrete Wavelet Transform to Model Whistle Contours for Dolphin Species Classification [†]

Paula Lopez-Otero [1,‡]* , Laura Docio-Fernandez [2,‡] and Antonio Cardenal-López [2,‡]

1   Information Retrieval Lab, CITIC, Universidade da Coruña, 15071 A Coruña, Spain
2   Multimedia Technology Group, atlanTTic Research Center, Universidade de Vigo, E36310 Vigo, Spain;
     ldocio@gts.uvigo.es (L.D.-F.); cardenal@gts.uvigo.es (A.C.-L.)
*   Correspondence: paula.lopez.otero@udc.gal; Tel.: +34-981-670000 (ext. 1276)
†   Presented at the XoveTIC Congress, A Coruña, Spain, 27–28 September 2018.
‡   These authors contributed equally to this work.

Published: 17 September 2018

check for
updates

Abstract: This work proposes the use of features based on the discrete wavelet transform (DWT) for dolphin species classification. These features are compared with other previously used in the literature, and the experiments carried out in a database featuring four different species of cetaceans (three dolphins and a pilot whale) showed that the use of DWT features led to improved classification performance.

Keywords: cetacean classification; whistle contour; discrete wavelet transform

1. Introduction

The health of marine mammal populations is often considered an indicator of overall marine ecosystem health and resilience [1], which makes it interesting to develop automatic tools to detect and classify cetacean species. Cetaceans produce characteristic sounds, such as whistles and clicks, that are used for different tasks such as navigation, communication or hunting [2]. The whistles are short narrow bandwidth sounds of short duration, and they consist in omnidirectional tones that vary with time, sometimes presenting a strong harmonic structure. In addition, whistle patterns vary between species due to physiological differences or environmental conditions, among others [3], making it possible to distinguish cetacean species using these sounds.

Several efforts for classifying cetacean species using whistle contours have been made. In [4] and [5], different parameters extracted from whistles were used to train a species classifier. Other features were explored in [6], such as the number of harmonics and frequency ratios. A statistical analysis of the discrimination capability of these features for classifying some dolphin species can be found in [7]. Other researchers used cepstral coefficients (CCs) for this task [8]. With respect to the classification strategy, some researchers rely on classifiers such as discriminant function analysis or classification and regression trees [3], while the use of Gaussian mixture models (GMMs) is also common [8,9]. In this work, a set of statistics obtained from the discrete wavelet transform of whistle contours is proposed for cetacean species classification.

2. Materials and Methods

In this work, four different cetacean species are considered: common dolphin *Delphinus delphis* (DDE), striped dolphin *Stenella coeruleoalba* (SCO), common bottlenose dolphin *Tursiops truncatus* (TTR) and long-finned pilot whale *Globicephala melas* (GME). A set of whistles was extracted as proposed in [9], which led to 24, 15, 15 and 23 whistles of DDE, GME, SCO and TTR, respectively. These whistles were extracted from a set of recordings collected by CEMMA (http://www.cemma.org/) (Indemares

Life 07/NAT/E/000732 project (http://www.indemares.es)) in the Northwest of Spain (Galician Bank and Avilés Canyon) in 2010 and 2011.

The use of the discrete wavelet transform (DWT) for whistle characterization is proposed in this paper. The DWT is a multi-resolution technique that decomposes a time series into subsequences at different resolution scales providing data into high and low-frequency components. At high frequency, the wavelets can capture discontinuities, ruptures and singularities in the original data. At low frequency, the wavelet characterizes the coarse structure of the data to identify the long-term trends. Thus, the wavelet analysis allows to extract the hidden and significant temporal features of the original data. The first step consists in decomposing the original signal into approximation (CA) and detail (CD) coefficients by convolving the signal with a low-pass filter (LP) and a high-pass filter (HP), respectively. The low-pass filtered signal is the input for the next iteration level and so on. The approximation coefficients contain the general trend (the low-frequency components) of the signal, and the detail coefficients contain its local variations (the high-frequency components).

The analysis of whistle contours using DWT was carried out as follows: first the whistle contours were processed with a DWT (Daubechies-5 wavelet) and both approximation (*a*) and detail (*d*) coefficients were extracted from these raw contours and its logarithm version in four levels. Then, several features were obtained: percentage of energy of *a* and *d* coefficients in each level; relative wavelet energy and wavelet entropy; Shannon energy entropy and log-energy entropy of *a* and *d* in each level; RMS and standard deviation of *a* and *d* in each level; mean, standard deviation, skewness and kurtosis of the Teager-Kaiser Energy Operator (TKEO) [10] of *a* and *d* in each level.

3. Results

Four different systems were assessed:

- CC. Cepstral coefficients combined with a likelihood classifier based on GMMs as proposed in [8]. CCs were computed using 51 filters separated 500 Hz from each other, and uniformly distributed between 4 KHz and 30 KHz. The features were extracted every 1 ms using a 7 ms window.
- Whistle countour. Contour estimation consists in estimating the exact frequency of each whistle, aiming at obtaining an accurate detection of all the whistles in a signal. The frequency of whistle contours can be extracted using the unpredictability measure described in [9]. As in the case of CCs, the classification stage using whistle contours can be carried out by means of maximum posterior probability computation, as in [9].
- Whistle contour statistics (similarly to [3]): beginning and end frequencies, minimum and maximum frequencies, duration, slope of the beginning and end sweeps, number of inflection points, number of steps, frequency range. Support vector machine classifier with Gaussian kernel was used to perform species classification.
- Discrete wavelet transform as described in Section 2.

Table 1 shows the results of cetacean classification using the aforementioned systems following a leave-one-out strategy (the reported accuracies were obtained with the optimal parameters of each classifier). The outstanding performance of CCs for this classification task was proven to be misleading since the features are capturing information relative to the channel: this was demonstrated by means of a location-independent experiment, where these features showed an accuracy of 33.33%. In addition, an analysis of the confusion matrices of the different systems showed that DWT features were the only ones that achieved an accuracy above 50% for all classes.

Table 1. Accuracies obtained in the cetacean species classification experiment using four different systems.

Features	Accuracy
CC	86.59%
Whistle contour	55.84%
Whistle contour statistics	63.64%
Discrete wavelet transform	68.83%

4. Conclusions

This paper proposed the use of features extracted from the discrete wavelet transform for cetacean species classification using whistle contours. A comparison of these features with other approaches found in the literature showed an absolute improvement of 13% with respect to using the frequencies of the whistle contours and of 5% with respect to using a set of statistics extracted from the whistle contours. Nevertheless, the overall classification accuracy was around 69%, so there is still room for improvement.

Funding: This work has received financial support from (i) "Ministerio de Economía y Competitividad" of the Government of Spain and the European Regional Development Fund (ERDF) under the research projects TIN2015-64282-R and TEC2015-65345-P, (ii) Xunta de Galicia (projects GPC ED431B 2016/035 and GRC2014/024), and (iii) Xunta de Galicia—"Consellería de Cultura, Educación e Ordenación Universitaria" and the ERDF through the 2016–2019 accreditations ED431G/01 ("Centro singular de investigación de Galicia") and ED431G/04 ("Agrupación estratéxica consolidada").

Acknowledgments: The authors would like to thank CEMMA (Coordinadora para o Estudo dos Mamíferos Mariños) for making available all its underwater recordings and information on the studied marine mammals and their behaviour.

Conflicts of Interest: The authors declare no conflict of interest. The founding sponsors had no role in the design of the study; in the collection, analyses, or interpretation of data; in the writing of the manuscript, and in the decision to publish the results.

References

1. Frasier, K.; Roch, M.; Soldevilla, M.; Wiggins, S.; Garrison, L.; Hildebrand, J. Automated classification of dolphin echolocation click types from the Gulf of Mexico. *PLoS Comput. Biol.* **2017**, *13*, e1005823.
2. Au, W.; Popper, A.; Fay, R. Hearing by whales and dolphins. In *Springer Handbook of Auditory Research*; Springer: New York, NY, USA, 2000.
3. Oswald, J.; Barlow, J.; Norris, T. Acoustic identification of nine delphinid species in the eastern tropical Pacific Ocean. *Mar. Mamm. Sci.* **2003**, *19*, 20–37.
4. Oswald, J.; Rankin, S.; Barlow, J.; Lammers, M. A tool for real-time acoustic species identification of delphinid whistles. *J. Acoust. Soc. Am.* **2007**, *122*, 587–595.
5. López, B.D. Whistle characteristics in free-ranging bottlenose dolphins (Tursiops truncatus) in the Mediterranean Sea: Influence of behaviour. *Mamm. Biol. Z. Säugetierkd.* **2011**, *76*, 180–189.
6. Soto, A.B.; Marsh, H.; Everingham, Y.; Smith, J.; Parra, G.; Noad, M. Discriminating between the vocalizations of Indo-Pacific humpback and Australian snubfin dolphins in Queensland, Australia. *J. Acoust. Soc. Am.* **2014**, *136*, 930–938.
7. Gannier, A.; Fuchs, S.; Quèbre, P.; Oswald, J. Performance of a countour-based classification method for whistles of Mediterranean delphinids. *Appl. Acoust.* **2002**, *71*, 1063–1069.
8. Roch, M.; Soldevilla, M.; Burtenshaw, J.; Henderson, E.; Hildebrand, J. Gaussian mixture model classification of odontocetes in the Southern California Bight and the Gulf of California. *J. Acoust. Soc. Am.* **2007**, *121*, 1737–1748.
9. Parada, P.P.; Cardenal-López, A. Using Gaussian mixture models to detect and classify dolphin whistles and pulses. *J. Acoust. Soc. Am.* **2014**, *135*, 3371–3380.
10. Kaiser, J. Some Useful Properties of Teager's Energy Operators. In Proceedings of the 1993 IEEE International Conference on Acoustics, Speech, and Signal Processing, Minneapolis, MN, USA, 27–30 April 1993; Volume III, pp. 149–152.

proceedings

MDPI

Extended Abstract

Automation of the Data Acquisition System for Self-Quantification Devices [†]

Rodrigo Martín-Prieto

RNASA-IMEDIR Research Group, University of A Coruña, A Coruña 15001, Spain; r.martin1@udc.es;
Tel.: +34-981-167-000

[†] Presented at the XoveTIC Congress, A Coruña, Spain, 27–28 September 2018.

Published: 19 September 2018

Abstract: In this paper we describe an environment that enables the interaction and data-fetching through a computer system from the Xiaomi Mi Band 2, a very popular and inexpensive Bluetooth Low Energy Fitness device, thus making it suitable for health-care long-term projects in which continuously gathering sleep and activity data is required. The environment is composed by a communication server running a custom library, exposed through both a shell command prompt and a RESTful API so it can be used by any other system connected to this server. The library is capable of connecting to an arbitrary number of devices at a time, bypassing many restrictions of the manufacturer's proprietary application and not depending on an outgoing network connection to synchronize data between the system and the devices. In this paper we cover not only the process to enable communication with the target device from computers but also the architectural aspects of the developed system. We also provide brief information about the prototypes developed to test the system on a real ongoing geriatric study.

Keywords: wearables; reverse-engineering; Bluetooth; Ble; fitness; clinical studies

1. Introduction

Bluetooth Low Energy [1] (BLE henceforth) wearable fitness devices have become more and more ubiquitous and inexpensive in the later years [2]. These devices have become great and inexpensive tools to conduct clinical studies due to their capability to gather daily activity and sleep data without limiting the study subjects' freedom [3].

Non clinical-oriented devices are intended for a rather individual use, making the user pair a single device to it's personal *smartphone* instead of being able to collect the raw data from multiple wearable devices from a computer.

Despite the limitations imposed by commercial lower-end wearable devices, they are already being used in experiments in fields such as pediatrics and gerontology with satisfying results [4]. The main issue is that these limitations impose a cap to how many subjects can take part in an experiment as working with one-to-one pairing and without having methods to automatically extract the required data, this has to be done manually.

2. Objectives

The main objective is to reverse engineer an inexpensive, common and lower-end fitness device, the Mi Band 2 by the Chinese manufacturer XiaoMi and develop a library that is able to seamlessly connect to this kind of device and gather data from many of them at a time to enable these devices for research use. Associated to this library there will be some other tools to expose its core functionality and be able to develop software on top of it. Namely, these tools would be:

1. **Shell:** A command line to interact with the surrounding devices and command the computer to fetch the raw data from them
2. **REST API:** A simple API to expose the funcionality of the library to other technologies and be able to integrate it in more complex projects.

3. Materials and Methods

The first stage of the project was getting to know the profile of the Mi Band 2 as a BLE Peripheral and the commands that need to be issued in order to connect, authenticate and fetch the activity data from the device. This was done through the use of Android development tools and a Bluetooth Sniffing device alongside with the official Mi Fit smartphone application to register the most important ones. After this, we were able to obtain a full specification of all the endpoints of the Mi Band 2 and at least the relevant commands and response codes to perform the functionality we were aiming for. These commands were later implemented in Python to create our own independent library with the aid of the Python module "bluePy" to avoid implementing low-level Bluetooth functionality. Mi Band 2 devices were modeled in a Python Class that inherited from bluePy's generic own BLE Peripheral Class and then the extracted BLE commands and responses were implemented as Class methods of our Mi Band 2 Class.

As for the library-related tools, the SHELL was developed alongside the library to test it's functionality and conduct early prototype tests, while the API was developed once the main methods for authentication, connection and activity data fetching worked as expected. The REST API was finally used to integrate the library with the GeriaTIC project through a module inside their own platform: ClepITO.

Initially, we proposed a fully distributed architecture to integrate the library with the ClepITO platform. The library would run inside a Data Automatic Acquisition Server in a Linux System and then would connect to the platform's own database to read and store data, while the ClepITO platform would control the Acquisition Server through HTTP requests making use of the REST API.

Due to some limitations of the project, we ended up using a fully integrated architecture, implementing the server inside a Linux virtual machine running in the same server as the ClepITO platform, and connecting both systems through a host-only network that provided bidirectional access to both systems, allowing for the same functionality as in a distributed environment but encapsulating everything under the same physical system. (See Figure 1)

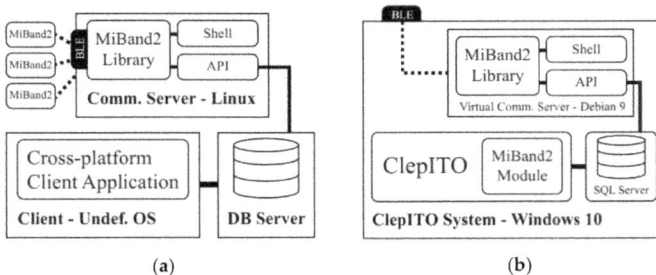

(a)　　　　　　　　　　　　　　(b)

Figure 1. (**a**) Initial design for the full system's architecture. Each element is encapsulated on it's own subsystem and interconnected by local or external networks for maximum flexibility. (**b**) Final design for the full system's architecture. All of the elements are packed under the same super-system due to project's requirements and for ease of replication purposes.

4. Results and Conclusions

To test the library's capabilities, two field tests were conducted during two visits to one of the centers associated with the GeriaTIC project located at Carballo, A Corunna. The first one was done early in the development of the library, making use of the Shell developed to test the library, while the second one made use of the REST API. Both tests resulted successful, and we were able to seamlessly gather data from all of the study participants. The first test took a few minutes to connect and fetch all of the data from the devices, but this time was greatly improved to less than a minute per device for the second test.

The data extracted by hand from the Mi Fit application is previously aggregated by the application's own algorithms, while being able to autonomously fetch the data from these devices provide us with a minute-by-minute data feed that we can manipulate and interpret at will, resulting in an improvement of the data quality. To conclude, we were able to develop a complete novel system capable of communicating a computer with a closed-source wearable device, the Mi Band 2, and improved the data gathering process in effort and time. By doing so, we have not only improved the methods and results of the GeriaTIC project but also have enabled other similar projects to perform better than before and it's our hope that we have encouraged them to use inexpensive wearable devices such as the Mi Band 2 to improve these studies.

This work is, in the end, a proof of concept about BLE devices and how we can communicate with them in ways that they were not initially meant to, and how these same devices can help in health-care areas. This research also manifests the need of more accessible and open wearable devices so researchers can conduct more and better studies without having to crack open closed devices for that mean. As both the Mi Band 2 and the Mi Fit application are closed-source, we lack the ability to know for sure if our implementation of the commands to communicate with the device is complete, or if our way of interpreting and aggregating the data is completely correct, so there is still work and tests that need to be done in order to assure that these are as good or better than the originals.

Acknowledgments: Thanks to Eloy Naveira [5] for his little, but meaningful speech about his undergraduate thesis on opening the Mi Band 1 during the master course I attended to. This speech was the last push I needed to start working on opening the Mi Band 2. Also thanks to the FreeYourGadget community for providing with their constant knowledge and the source of the GadgetBridge Android application, that helped me understand the inner workings of the Mi Band 2. Last but not least, thanks to Javier Pereira and the RNASA-IMEDIR laboratory, as well as the CITIC, for supporting my research both materially and economically. Without this support it would have been extremely difficult to conduct all of these tasks and take this research to a good end.

Conflicts of Interest: The author declares no conflict of interest. The founding sponsors had no role in the design of the study; in the collection, analyses, or interpretation of data; in the writing of the manuscript, and in the decision to publish the results.

References

1. Gómez: C.; Oller, J.; Paradells, J. Overview and Evaluation of Bluetooth Low Energy: An Emerging Low-Power Wireless Technology. *Sensors* **2012**, *12*, 11734–11753. doi:10.3390/s12091173.
2. Wearable Device Shipments Worldwide from 2016 to 2022. Available online: https://www.statista.com/statistics/610478/wearable-device-shipments-worldwide (accessed on 20 June 2018).
3. Technologies for Participatory Medicine and Health Promotion in the Elderly Population (GERIATIC). Available online: https://clinicaltrials.gov/ct2/show/study/NCT03504813?cntry=ES (accessed on 20 June 2018).
4. Saner, H. Wearable Sensors for Assisted Living in Elderly People. *Front. ICT* **2018**, *5*, 1. doi:10.3390/s12091173.
5. Naveira, E. Medical Sensors to Prevent Sedentarism. Undergraduate Thesis, University of A Coruña, A Coruña, Spain, September 2016. (In Spanish)

proceedings

MDPI

Extended Abstract

Testing Goodness-of-Fit of Parametric Spatial Trends [†]

Andrea Meilán-Vila [1,*], Jean Opsomer [2], Mario Francisco-Fernández [1] and Rosa M. Crujeiras [3]

[1] Departamento de Matemáticas, Universidade da Coruña, 15071 A Coruña, Spain; mariofr@udc.es
[2] Westat Inc., Rockville, MD 20850, USA; JeanOpsomer@westat.com
[3] Departamento de Estadística, Análisis Matemático y Optimización, Universidade de Santiago de Compostela, 15782 Santiago de Compostela, Spain; rosa.crujeiras@usc.es
* Correspondence: andrea.meilan@udc.es
† Presented at the XoveTIC Congress, A Coruña, Spain, 27–28 September 2018.

Published: 17 September 2018

check for
updates

Abstract: The aim of this work is to propose and analyze the behavior of a test statistic to assess a parametric trend surface, that is, a regression model with spatially correlated errors. The asymptotic behavior under the null hypothesis, as well as the asymptotic power of the test under local alternatives will be analyzed. Finite sample performance of the test is addressed by simulation, introducing a bootstrap calibration procedure.

Keywords: model checking; spatial trend; local linear regression; least squares; bootstrap

1. Introduction

Consider a spatial stochastic process, which consists of a collection of random variables indexed on a certain domain of $\mathbb{R}^2$, with a well-defined joint distribution. In this framework, the observed data usually exhibit an important feature: close observations tend to be more similar than those which are far apart. Therefore, such observations cannot be treated as independent and the dependence structure should be taken into account in any descriptive or inferential procedure. In particular, from the perspective of spatial regression models (a trend surface plus an error term), the dependence structure should be considered and properly introduced into the model.

A common task in statistics is to determine whether a parametric model is an appropriate representation of a dataset. Under the assumption of independent errors, some authors have developed goodness-of-fit tests for parametric models that rely on a smooth alternative estimated by a nonparametric regression method, as [1] or [2].

A new proposal for testing a parametric trend surface is given in this paper. The proposed test is based on a comparison between a smooth version of a parametric fit with a nonparametric estimator of the trend (specifically, the multivariate local linear estimator will be used) in terms of a distance.

2. Statistical Model

Let $\{Z(\mathbf{s}), \mathbf{s} \in D\}$ be a random spatial process consisting of collections of random variables indexed in a domain $D \subset \mathbb{R}^2$ with a well-defined joint distribution. Consider n locations $\{\mathbf{s}_1, \ldots, \mathbf{s}_n\}$ on the region D generated from a density f. The set of random variables corresponding with those locations will be represented by $\{Z(\mathbf{s}_1), \ldots, Z(\mathbf{s}_n)\}$. Assume the model

$$Z(\mathbf{s}_i) = m(\mathbf{s}_i) + \varepsilon(\mathbf{s}_i), \quad i = 1, \ldots, n, \tag{1}$$

where m is an unknown smooth regression function which is supposed to be twice continuously differentiable. The ε are unobserved random variables with

$$\mathbb{E}[\varepsilon(\mathbf{s}_i)] = 0, \quad \text{Cov}(\varepsilon(\mathbf{s}_i), \varepsilon(\mathbf{s}_j)) = \sigma^2 \rho_n(\mathbf{s}_i - \mathbf{s}_j), \quad i, j = 1, \ldots, n,$$

where $\sigma^2 < \infty$ and ρ_n is a continuous correlation function satisfying $\rho_n(0) = 1$, $\rho_n(\mathbf{s}) = \rho_n(-\mathbf{s})$ and $|\rho_n(\mathbf{s})| \leq 1$, $\forall \mathbf{s}$. The goal of this work is to test if the trend function belongs to a parametric family:

$$H_0 : m \in \mathcal{M}_\beta = \{m_\beta, \beta \in \mathcal{B}\}, \qquad \text{vs.} \qquad H_a : m \notin \mathcal{M}_\beta, \tag{2}$$

with $\mathcal{B} \subset \mathbb{R}^p$ a compact set. One of the more usual approaches is to compare a smooth version of a parametric fit with a nonparametric estimator of $m(\mathbf{s})$ and "thereafter" to reject H_0 if the distance between both fits exceeds a critical value.

3. Test Statistic

A suitable test statistic in order to solve the testing problem (2) could be computed as a weighted L_2—distance between the nonparametric and parametric fits, as in [2]:

$$T_n = n|\mathbf{H}|^{1/2} \int_D (\hat{m}_{\mathbf{H}}^{LL}(\mathbf{s}) - \hat{m}_{\mathbf{H}, \hat{\beta}}^{LL}(\mathbf{s}))^2 w(\mathbf{s}) d\mathbf{s}, \tag{3}$$

where w is a weight function. A full definition of the elements of the test statistic T_n can be found in Appendix A. For the calibration of the critical values, a bootstrap procedure is considered, see Appendix B.

4. Simulations

In this section, a simulation study showing the performance of the bootstrap procedure is presented. For this purpose, 500 samples of size $n = 400$ are generated from an isotropic spatial process observed at regularly spaced locations $\{\mathbf{s}_1, \ldots, \mathbf{s}_n\}$ in the unit square, where $\mathbf{s}_i = (s_{i1}, s_{i2})$, $i = 1, \ldots, n$:

$$Z(\mathbf{s}_i) = 2 + s_{i1} + s_{i2} + cs_{i1}^3 + \varepsilon(\mathbf{s}_i), \quad 1 \leq i \leq n. \tag{4}$$

The random errors $\varepsilon(\mathbf{s}_i)$ are normally distributed with zero mean and exponential covariance function $\text{Cov}(\varepsilon(\mathbf{s}_i), \varepsilon(\mathbf{s}_j)) = \sigma^2 \{\exp(-\|\mathbf{s}_i - \mathbf{s}_j\|/a_e)\}$, with $\sigma = 0.4$ and $\sigma = 0.8$. Different values of parameter a_e are considered: $a_e = 0.4, 0.6, 0.8$. The bootstrap procedure has been performed using $B = 500$ replicas for each sample. The weight function used was taken as $w(\mathbf{s}) = 1$. For simplicity, the bandwidth matrix was considered $\mathbf{H} = \text{diag}(h, h)$, and different bandwidth values were chosen, $h = 0.10, 0.15, 0.20$.

In Table 1, the simulated rejection probabilities obtained for T_n are presented for the significance level $\alpha = 0.05$ over the 500 trials. When c is equal to zero (under the null hypothesis of linearity of the trend), the proportion of rejections obtained is similar to the considered significance level, but this proportion depends directly on the value of the bandwidth h. When c is equal to 5 or 10, the power of the test is really good, since the proportion of rejections is close to one, in the majority of the cases. Again, this proportion depends on the value of the bandwidth.

Table 1. Proportion of rejections of the null hypothesis.

				h	
σ	a_e	c	0.10	0.15	0.20
0.4	0.4	0	0.052	0.047	0.042
		5	0.897	0.932	0.911
		10	0.905	0.948	0.923
0.4	0.6	0	0.054	0.042	0.034
		5	0.856	0.901	0.898
		10	0.894	0.926	0.918
0.8	0.8	0	0.068	0.048	0.038
		5	0.808	0.798	0.806
		10	0.845	0.803	0.816

Funding: This research has received financial support from the Xunta de Galicia and the European Union (European Social Fund-ESF). This research has been partially supported by MINECO grant MTM2014-52876-R, MTM2016-76969-P and MTM2017-82724-R and by the Xunta de Galicia (Grupos de Referencia Competitiva ED431C-2016-015 and Centro Singular de Investigación de Galicia ED431G/01), all of them through the ERDF.

Appendix A

The trend surface estimation can be performed using a parametric and a non-parametric approach. In the parametric context, an iterative estimation procedure could be used. Denoting $\mathbf{Z} = (Z(\mathbf{s}_1), \cdots, Z(\mathbf{s}_n))'$ and $\mathbf{m}_\beta = (m_\beta(\mathbf{s}_1), \ldots, m_\beta(\mathbf{s}_n))'$, under H_0 the steps of the procedure are:

(1) Based on the sample, estimate the trend parameter β using the ordinary least squares estimator, ignoring the dependence structure of the errors:

$$\tilde{\beta} = \arg\min_\beta (\mathbf{Z} - \mathbf{m}_\beta)'(\mathbf{Z} - \mathbf{m}_\beta).$$

(2) Estimate the variance-covariance matrix of the errors Σ using the residuals $\tilde{\varepsilon}(\mathbf{s}_i) = Z(\mathbf{s}_i) - m_{\tilde{\beta}}(\mathbf{s}_i)$, $i = 1, \ldots, n$, obtained from the estimator of the trend from Step (1). Note that, the entries of Σ are:

$$\Sigma(i, j) = C_\theta(\mathbf{s}_i - \mathbf{s}_j), \quad i, j = 1 \ldots, n,$$

where $C_\theta(\mathbf{s}_i - \mathbf{s}_j) = \sigma^2 - \gamma_\theta(\mathbf{s}_i - \mathbf{s}_j)$, being $\{2\gamma_\theta(\mathbf{u}) : \theta \in \Theta \subset \mathbb{R}^q\}$ a valid parametric family to estimate the variogram function.

(3) Estimate the trend parameter β using the weighted least squares estimator, taking the dependence structure of the errors into account:

$$\hat{\beta} = \arg\min_\beta (\mathbf{Z} - \mathbf{m}_\beta)'\tilde{\Sigma}^{-1}(\mathbf{Z} - \mathbf{m}_\beta).$$

Therefore, the parametric trend estimator considered would be $m_{\hat{\beta}}$. Note that, an estimation of Σ can be obtained from the residuals $\tilde{\varepsilon}(\mathbf{s}_i)$, $i = 1, \ldots, n$, as follows:

$$\tilde{\Sigma}(i, j) = C_{\tilde{\theta}_{LS}}(\mathbf{s}_i - \mathbf{s}_j) = \tilde{\sigma}^2 - \gamma_{\tilde{\theta}_{LS}}(\mathbf{s}_i - \mathbf{s}_j), \quad i, j = 1 \ldots, n,$$

where $\gamma_{\tilde{\theta}_{LS}}$ is the parametric least squares estimator of the variogram and $\tilde{\sigma}^2$ is an estimator of the variance. The last estimator could be obtained using a least squares procedure.

From a nonparametric point of view, model (1) has been studied by several authors. Some approaches used for this task include kernel-based methods. In this case, the trend is estimated using the multivariate local linear estimator, see [3]. In the spatial framework, the local linear estimator for $m(\mathbf{s})$ at a location $\mathbf{s}$ can be explicitly written as

$$\hat{m}_{\mathbf{H}}^{LL}(\mathbf{s}) = \mathbf{e}_1'(X_s'W_sX_s)^{-1}X_s'W_s\mathbf{Z},$$

where $\mathbf{e}_1 = (1,0,0)'$, X_s is a $n \times 3$ matrix whose i-th row equals $(1,(\mathbf{s}_i - \mathbf{s})')$, $i = 1,\ldots,n$, $W_s = \text{diag}\{K_{\mathbf{H}}(\mathbf{s}_1 - \mathbf{s}),\ldots,K_{\mathbf{H}}(\mathbf{s}_n - \mathbf{s})\}$, where $K_{\mathbf{H}}(\mathbf{s}) = |\mathbf{H}|^{-1}K(\mathbf{H}^{-1}\mathbf{s})$ is used to assign weights. $\mathbf{H}$ is a 2×2 symmetric, positive definite matrix depending on the sample size n and K is a multivariate kernel function. Given $\mathbf{s}$, the bandwidth $\mathbf{H}$ controls the shape and the size of the local neighborhood used to estimate m.

Therefore, taking into account these estimators, the proposed test statistic is

$$T_n = n|\mathbf{H}|^{1/2} \int_D (\hat{m}_{\mathbf{H}}^{LL}(\mathbf{s}) - \hat{m}_{\mathbf{H},\hat{\beta}}^{LL}(\mathbf{s}))^2 w(\mathbf{s})d\mathbf{s},$$

where w is a weight function and $\hat{m}_{\mathbf{H},\hat{\beta}}^{LL}$ is a smooth version of the parametric estimator $m_{\hat{\beta}}$, which is defined by

$$\hat{m}_{\mathbf{H},\hat{\beta}}^{LL}(\mathbf{s}) = \mathbf{e}_1'(X_s'W_sX_s)^{-1}X_s'W_s\mathbf{m}_{\hat{\beta}},$$

where $\mathbf{m}_{\hat{\beta}} = (m_{\hat{\beta}}(\mathbf{s}_1),\ldots,m_{\hat{\beta}}(\mathbf{s}_n))'$.

Appendix B

Once a suitable test statistic is available, a crucial task is the calibration of critical values for a given level α, namely t_α. Usually, the estimation of these critical values t_α such that $\mathbb{P}_{H_0}(T_n \geq t_\alpha) = \alpha$ can be done by means of the asymptotic distribution. The use of asymptotic theory to calibrate the test poses some problems, such as the need to estimate some nuisance functions and a slow convergence rate to the limit distribution. Under these circumstances, calibration can be done by means of resampling procedures, such as bootstrap, see [4].

The procedure consists in generating a bootstrap sample $\{Z^*(\mathbf{s}_i), i = 1,\ldots,n\}$ and then computing a bootstrap statistic T_n^* like T_n by the squared deviation between the smooth version of the parametric fit $\hat{m}_{\hat{\beta}^*}^{LL}$ and the nonparametric fit $\hat{m}^{*LL}$. Once the bootstrap statistic is computed, the distribution of T_n^* can be approximated by Monte Carlo. From this Monte Carlo approximation, the $(1 - \alpha)$ quantile t_α^* is defined and the parametric hypothesis es rejected if $T_n > t_\alpha^*$. The specific steps for the algorithm used in this work are the following:

1. Obtain the parametric trend estimator $\hat{\beta}$.
2. Estimate the covariance matrix of the errors $\hat{\Sigma}$ based on the residuals $\hat{\varepsilon} = (\hat{\varepsilon}(\mathbf{s}_1),\ldots,\hat{\varepsilon}(\mathbf{s}_n))'$, where $\hat{\varepsilon}(\mathbf{s}_i) = Z(\mathbf{s}_i) - m_{\hat{\beta}}(\mathbf{s}_i)$, $i = 1,\ldots,n$, and find the matrix L, such that $\hat{\Sigma} = LL'$, using Cholesky decomposition.
3. Compute the independent residuals, $\mathbf{e} = (e(\mathbf{s}_1),\ldots,e(\mathbf{s}_n))'$, given by $e(\mathbf{s}_i) = L^{-1}\hat{\varepsilon}(\mathbf{s}_i)$.
4. These independent variables are centered and, from them, we obtain an independent bootstrap sample of size n, denoted by $\mathbf{e}^* = (e^*(\mathbf{s}_1),\ldots,e^*(\mathbf{s}_n))$.
5. Finally, the bootstrap errors $\varepsilon^* = (\varepsilon^*(\mathbf{s}_1),\ldots,\varepsilon^*(\mathbf{s}_n))$ are $\varepsilon^*(\mathbf{s}_i) = Le^*(\mathbf{s}_i)$, and the bootstrap samples are $Z^*(\mathbf{s}_i) = m_{\hat{\beta}}(\mathbf{s}_i) + \varepsilon^*(\mathbf{s}_i)$.

References

1. Hardle, W.; Mammen, E. Comparing nonparametric versus parametric regression fits. *Ann. Stat.* **1993**, *21*, 1926–1947.
2. Alcalá, J.; Cristóbal, J.; González-Manteiga, W. Goodness-of-fit test for linear models based on local polynomials. *Stat. Probab. Lett.* **1999**, *42*, 39–46.

3. Fan, J.; Gijbels, I. *Local Polynomial Modelling and Its Applications: Monographs on Statistics and Applied Probability*; CRC Press: Boca Raton, FL, USA, 1996; Volume 66.
4. Francisco-Fernández, M.; Jurado-Expósito, M.; Opsomer, J.; López-Granados, F. A nonparametric analysis of the spatial distribution of Convolvulus arvensis in wheat-sunflower rotations. *Environmetrics* **2006**, *17*, 849–860.

proceedings

MDPI

Extended Abstract

A Task Planning Problem in a Home Care Business [†]

Isabel Méndez Fernandez *, Ignacio García Jurado, Silvia Lorenzo Freire,
Luisa Carpente Rodríguez and Julián Costa Bouzas

Grupo MODES, Departamento de Matemáticas, Universidade da Coruña, 15071 A Coruña, Spain;
igjurado@udc.es (I.G.J.); slorenzo@udc.es (S.L.F.); luisacar@udc.es (L.C.R.); julian.costa@udc.es (J.C.B.)
* Correspondence: isabel.mendez.fernandez@udc.es; Tel.: +34-981-167-000
† Presented at the XoveTIC Congress, A Coruña, Spain, 27–28 September 2018.

Published: 19 September 2018

Abstract: This work focuses on the study of a task planning problem in a home care business. The objective is to schedule the working days of the available nurses, in order to assist all the active clients. Due to the large size of the real cases that must be faced, it is not possible to obtain exact solutions of the problem in short periods of time. Therefore, we propose an algorithm, which is based on heuristic techniques, to provide approximated solutions to the incidents that arise daily in the company. The designed algorithm is validated by obtaining the automatic schedule to solve a battery of real-like examples.

Keywords: optimization; scheduling; heuristic algorithms; operations research

---

1. Introduction

Home health care is a resource that allows elderly and/or dependent people to continue living at their homes despite being in a situation of dependency or in need of assistance to carry out different tasks of their daily lives. The company we are working with has been providing home care services since 1997, both in the city of A Coruña and in the neighboring municipalities. This company faces a scheduling problem, in which we have to determinate the routes the employees must follow and set the exact time at which the caregivers must perform the services assigned to them.

1.1. Concepts

The users of the service provided by the company are those who need health care support or require help in carrying out certain tasks and, to improve their quality of life, hire the company's services. These users must specify the number of services they require, detailing the day and the time frame (morning, noon, afternoon or evening) within which they must be completed. Also, they should indicate the available (when the service must be performed) and optimal (when the user prefers to be attended) time windows of each service.

To provide its services the company has a set of caregivers, which may increase or decrease depending on the volume of work available. These caregivers are responsible for visiting users' homes (at the time they have been instructed to do so) and carrying out the tasks assigned to them (for example: cooking meals, monitoring medication, cleaning chores, etc.). For each caregiver and user, the company sets a level of affinity, which establishes how suitable the caregiver is to assist the user.

Therefore, the company must carry out a work plan where, for each caregiver, the services to be provided are specified (establishing the tasks to be performed in each of them) and the times at which those services must be completed (always upholding the availability time windows established by the users). This work plan arranges the whole week, so the schedule is repeated over time until the need to modify it arises. In order to carry out this task, the company has several organizers who are in charge of managing the working days, for each caregiver, manually. Each organizer works with a set of users and assistants, according to the areas or neighborhoods he/she operates in.

1.2. Objective

The company's objective is to obtain a computer tool that automatically modifies the caregivers' previous plan in order to solve the following incidents: (i) registration of a user, (ii) discharging a user, (iii) increasing the number of services required by a user, (iv) decreasing the number of services required by a user, (v) alteration of any of the parameters of a service, (vi) repetition and/or combination of the previous instances.

To obtain the new schedules, the company asked us to consider the following objectives:

1. The new plans should minimize time lost between services.
2. The new plans should maximize the affinity levels between users and the caregivers that visit them.
3. We must try not to change the previous planning in excess.
4. The new plans should respect, as far as possible, the optimal time windows of the services.

The purpose of this work is to provide the company, employing different operational research techniques, with a computer tool that allows them to automatically modify the caregivers' schedules in order to solve the abovementioned incidents.

2. Problem Solution

The problem presented by the company has been modeled as an integer linear programming problem. For real cases, such as those faced daily in the company, it presents a large number of variables and constraints. Therefore, it is not always possible to obtain the optimal schedule by solving the linear programming problem.

Because of this, the problem must be solved in an approximated way, through the use of metaheuristic techniques, in order to obtain admissible solutions in short computational times.

Algorithm

The algorithm developed to solve the company's planning problem is based on the simulated annealing method. This algorithm consists of re-scheduling the services and optimizing the working days of the caregivers.

The algorithm is designed to solve the incidents the company works with, avoiding excessive modification of the previous planning. The aim we follow is to make the work plans feasible while trying to ensure that the caregivers have no gaps in their working hours. The most important steps of the algorithm are presented below:

1. Preparation of elements: The algorithm starts with an initial solution, which is the previous schedule assigned to the caregivers. The services that need to be removed from this solution are deleted and, at the same time, we select the services that need to be rescheduled and the ones that need to be included in the plans.
2. Service planning: Each of the services that must be planned is assigned to the best caregiver available. After that the service is scheduled in such a way that causes the least amount of overlap and break gaps (between the new one and the rest of services assigned to said caregiver). To establish how suitable a caregiver is to assist a user, the following aspects are considered: (i) the affinity level between them, (ii) the difference between the hours worked by the assistant and those specified in her contract, and (iii) the travel time from that user's home to all the clients the caregiver is working with. In some cases, we don't have services to plan, so the algorithm goes directly to the optimization phase.
3. Optimization: After assigning a service to a caregiver, the schedule is optimized by applying the simulated annealing method to eliminate possible overlaps or breaks in the schedule. If, after the optimization, the previous overlaps have not been eliminated, the service must be assigned

to the next best caregiver available. In the cases where there are no services to be scheduled, the caregivers' plan is optimized to eliminate possible break gaps in it (in such circumstances there is no overlap between services).

4. End: When all the services have been correctly planned, the overlaps have been eliminated and the breaks have been optimized, the algorithm ends, returning the optimized planning as the new solution.

The movements we considered in the implementation of the simulated annealing method are: rescheduling one service, rescheduling several services at once, exchanging the time in which two selected services are carried out, exchanging the caregivers that conduct two selected services, and changing the caregiver that performs a service.

The objective function considered to evaluate a schedule in the simulated annealing method is a lexicographic function, in which we want to minimize the following aspects: (i) the overlap time between services, (ii) the time lost between services, and (iii) the time that the services are carried out outside their optimal time window.

3. Conclusions

In this work, we analyze the problem presented by the company and we design a resolution method based on the metaheuristic technique of the simulated annealing. The purpose of this algorithm is to solve the incidents considered by the company, trying to modify the original schedules as little as possible.

To check the algorithm's performance, we design a series of examples, in which we take into account all the different incidents that the company deals with, and we solve them using the method previously designed. Based on the results obtained, we can assume that the algorithm's behavior is acceptable, since it correctly solves all the incidents considered while using low computational times.

Author Contributions: I.M.F. implemented the algorithm and performed the experiments; I.M.F., I.G.J., S.L.F., L.C.R. and J.C.B. designed the algorithm.

Acknowledgments: This research was funded by the Ministerio de Economóa y Competitividad though the proyect MTM2014-53395-C3-1-P, by the Centro para el Desarrollo Tecnológico Industrial through the proyect ITC-20151247, and by the Xunta de Galicia through the aid for Grupos de Referencia Competitiva ED431C-2016-015 and Centro Singular de Investigación de Galicia ED431G/01.

Extended Abstract

Feature Selection with Limited Bit Depth Mutual Information for Embedded Systems †

Laura Morán-Fernández *, Verónica Bolón-Canedo and Amparo Alonso-Betanzos

CITIC, Universidade da Coruña, 15071 A Coruña, Spain; vbolon@udc.es (V.B.-C.); ciamparo@udc.es (A.A.-B.)
* Correspondence: laura.moranf@udc.es
† Presented at the XoveTIC Congress, A Coruña, Spain, 27–28 September 2018.

Published: 17 September 2018

Abstract: Data is growing at an unprecedented pace. With the variety, speed and volume of data flowing through networks and databases, newer approaches based on machine learning are required. But what is really big in Big Data? Should it depend on the numerical representation of the machine? Since portable embedded systems have been growing in importance, there is also increased interest in implementing machine learning algorithms with a limited number of bits. Not only learning, also feature selection, most of the times a mandatory preprocessing step in machine learning, is often constrained by the available computational resources. In this work, we consider mutual information—one of the most common measures of dependence used in feature selection algorithms—with reduced precision parameters.

Keywords: feature selection; mutual information; reduced precision; embedded systems; Big Data

---

1. Introduction

In the age of Big Data, with datasets being collected in almost all fields of human endeavor, there is an emerging economic and scientific need to extract useful information from it. Thus, machine learning algorithms have become indispensable. One machine learning technique is feature selection [1]. It arises from the need of determining the "best" subset of variables for a given problem. The use of an adequate feature selection method can avoid over-fitting and improve model performance, providing faster and more cost-effective learning models and a deeper insight into the underlying processes that generate the data. Features can be categorized in three ways: relevant, irrelevant and redundant. As a result, selecting the relevant features and ignoring the irrelevant and redundant ones is advisable.

The process of feature selection is typically performed on a machine using high numerical representation (64 bits). Using a more powerful processor provides significant benefits in terms of speed and capability to solve more complex problems. Although this capability does not come without cost; a conventional microprocessor can require a substantial amount of off-chip support hardware, memory, and often a complex operating system. In contrast to up-to-date computers, these requirements are often not met by embedded systems, low energy computers or integrated solutions that need to optimize the used hardware resources. With the power demand of smartphones, health wearables and fitness trackers, there is a need for tools that enable energy consumption estimation for such systems. Thus, we identify one such opportunity to develop a feature selection algorithm in embedded systems without reducing performance. This opportunity leverages the observation that algorithms yield parameters which can achieve performances close to that of optimal double-precision parameters by simply limiting the amount of bits. In this work, we investigate feature selection by considering the information theoretic measure of mutual information with reduced precision parameters. The mutual information measure is used due to its computational efficiency and simple interpretation. Therefore, we are able to provide a limited bit depth mutual information, and, through minimum Redundancy Maximum Relevance feature selection method, experimentally achieve classification performances close to that of 64-bit representations for several real and synthetic datasets.

2. Limited Bit Depth Mutual Information

In information theoretic feature selection, the main challenge is to estimate the mutual information [2]. To calculate mutual information we need to estimate the probability distributions. Internally, it counts the occurrences of values within a particular group. Thus, based on Tschiatschek's work [3] for approximately computing probabilities, we investigate mutual information with limited number of bits by considering this measure with reduced precision counters. To perform the reduced precision approach, we target a fixed-point representation instead of the 64-bit resolution.

Mutual Information parameters are typically represented in the logarithm domain. For the reduced precision parameters, we compute the number of occurrences and use a lookup table to determine the logarithm of the probability of a particular event. The lookup table is indexed in terms of number of occurrences of an event and the total number of events and stores values for the logarithms in the desired reduced precision representation. Following the fixed-point representation, and to limit the maximum size of the lookup table and the bit-width required for the counters, we assumed some maximum integer number. After calculating the cumulative count, in order to guarantee that the counts stay in range, the algorithm identifies counters that reach their maximum value, and halves these counters.

3. Experimental Results and Conclusions

Our limited depth mutual information can be applied to any method that uses internally the mutual information measure. We have chosen to do it within feature selection since with the advent of Big Data, feature selection process has a key role to play in helping reduce high-dimensionality in machine learning problems. There is a large number of feature selection methods that use mutual information as a measure, thus their performance depending on the accuracy obtained by the mutual information step. Among the different feature selection algorithms based on mutual information, the mRMR (minimum Redundancy Maximum Relevance) multivariate filter [4] is used due to its popularity and good results in the machine learning area.

Experimental results over several synthetic and real datasets have shown that 16 bits are sufficient to return the same feature ranking than that of double precision representation. Besides, classification results showed that even using a 4-bit representation, our limited bit depth mutual information was able to achieve performances very close to that of full precision mutual information. As a result, meaningful computational, runtime and memory benefits will be provided when implementing mutual information in embedded systems.

Acknowledgments: This research has been financially supported in part by the Spanish Ministerio de Economía y Competitividad (research project TIN2015-65069-C2-1-R), by European Union FEDER funds and by the Consellería de Industria of the Xunta de Galicia (research project GRC2014/035). Financial support from the Xunta de Galicia (Centro singular de investigación de Galicia accreditation 2016–2019) and the European Union (European Regional Development Fund–ERDF), is gratefully acknowledged (research project ED431G/01).

Conflicts of Interest: The author declares no conflict of interest.

References

1.   Guyon, I.; Elisseeff, A. An introduction to variable and feature selection. *J. Mach. Learn. Res.* **2003**, *3*, 1157–1182.
2.   Shannon, C.E. A mathematical theory of communication. *ACM SIGMOBILE Mob. Comp. Rev.* **2001**, 5, 3–55.

3. Tschiatschek, S.; Pernkopf, F. Parameter learning of Bayesian network classifiers under computational constraints. In Proceedings of the Joint European Conference on Machine Learning and Knowledge Discovery in Databases, Porto, Portugal, 7–11 September 2015; Springer, pp. 86–101.

4. Peng, H.; Long, F.; Ding, C. Feature selection based on mutual information criteria of max-dependency, max-relevance, and min-redundancy. *IEEE Trans. Pattern Anal.* **2005**, *27*, 1226–1238.

proceedings

MDPI

Extended Abstract

Promoting Active Aging and Quality of Life through Technological Devices †

Laura Nieto-Riveiro [1,2,3,*]**, Thais Pousada-García** [1,2,3] **and María del Carmen Miranda-Duro** [2,3,4]

[1] Health Sciences Department, Universidade da Coruña, 15071 A Coruña, Spain;
    thais.pousada.garcia@udc.es
[2] Research group Artificial Neural Networks and Adaptative, Systems-Center of Medical Informatics and
    Radiological Diagnosis, (RNASA-IMEDIR), Research Center on Information and Communication,
    Technologies (CITIC), Faculty of Health Sciences, Universidade da, Coruña, 15071 A Coruña, Spain;
    carmen.miranda@udc.es
[3] Institute for Biomedical Research (INIBIC), Universidade da Coruña, 15071 A Coruña, Spain
[4] Biomedical Sciences, Medicine and Physiotherapy Department, Universidade da Coruña, 15071 A Coruña,
    Spain
* Correspondence: laura.nieto@udc.es; Tel.: +34-981-167-000
† Presented at the XoveTIC Congress, A Coruña, Spain, 27–28 September 2018.

Published: 18 September 2018

Abstract: This abstract presents a set of projects developed by RNASA-IMEDIR research group of the Universidade da Coruña, aimed at promoting active aging, quality of life, health and personal autonomy of older people, by means of technological devices.

Keywords: active aging; healthy aging; information and communication technologies; older people; quality of life

1. Introduction

Active aging (AA) is defined, according to the World Health Organization [1] (p. 12), as 'the process of optimizing opportunities for health, participation, and security in order to enhance the quality of life as people age'. At the International Conference on Active Aging, held in Seville in 2010, a fourth component was included: lifelong learning.

In recent decades, several initiatives have been developed, aimed at promoting AA, some of them through technological tools, such as the 'Living with vitality' program [2,3]. In this regard, the literature shows that information technology and communications (ICT) can provide numerous benefits to the elderly, associated with different components of AA: stimulation or maintenance of cognitive abilities; opportunities for communication and socialization; participation in leisure activities; online use of administrative or financial services; purchases at a distance, or access to community resources such as health or social services [4–6].

This abstract presents a set of projects developed by RNASA-IMEDIR research group of the Universidade da Coruña, aimed at promoting AA, quality of life, health and personal autonomy of older people, by means of technological devices.

2. In-TIC Project

This work focuses on digital literacy and therapeutic stimulation of the older population through technological tools, using In-TIC PC software as support [7,8].

In-TIC PC is a software created by RNASA-IMEDIR group, with the support of Orange Foundation. It's available for free download on the page www.intic.udc.es. It allows to simplify and personalize the use of the computer and the Internet, by configuring virtual keyboards adapted to the physical, cognitive, sensory and social needs of each person. In this way, a set of virtual keyboards, specifically adapted to the needs and capacities of the 10 elderly participants in the project, was prepared to facilitate their access to ICT and their digital literacy process.

In addition, a teaching methodology, based on the learning interests and rhythms of the older people, was elaborated, that includes activities and contents related to the application of ICT in their daily life, leisure and social participation.

3. EA-TIC Project

EA-TIC is a project based on the use of an interactive digital whiteboard (IDW) as a means of intervention [9]. Through this technological support, two programs of different activities were implemented with 45 older participants, divided into groups of 3–10 persons.

- Security and Healthy Living Program, aimed at reinforcing the safety of the older people in their occupational performance, both in their homes and in other habitual environments, and promoting the acquisition of healthy living habits.
- Leisure and Social Participation Program, aimed at favoring their social participation and their participation in meaningful leisure activities.

The results obtained in this project showed that the participants had positive impressions on the IDW, highlighting different advantages provided by this tool: accessibility of content, related to the wide projection surface; speed and dynamism in the presentation of content; immediacy in finding information; possibility to capture on the board the contributions of the group's members; stimulation of the capacity of attention and concentration; or support in structuring and understanding information.

4. CloudPatient Project

CloudPatient is a project focused on the use of a platform in the cloud that allows interoperability between biomedical sensors, mobile devices and cloud services for intelligent attention to the older people (www.cloudpatient.udc.es). This platform facilitates the monitoring of their treatments and health status, and offers digital contents and activities, aimed at promoting a healthy and safe lifestyle.

The main services offered by the CloudPatient platform are:

- Semi-automatic registry, through biosensors and wearables, of different parameters related to the health status of the person.
- Monitoring of certain factors related to health status and well-being, such as physical activity, sleep quality, weight or blood pressure.
- Generation of personalized multimedia contents and activities, focused on promoting a healthy and safe lifestyle and improving their quality of life.
- Access to updated, unified, accessible, and reliable information in real time.

The pilot tests of this project were carried out with 10 elderly people. For several months, they used the CloudPatient platform and biosensors, as a wearable to record physical activity and sleep quality.

5. Scratch-EA Project

The general objective of Scratch-EA project is to encourage creativity development, to improve quality of life and to promote AA, through the use of Scratch, a computer programming tool.

The intervention developed in this project lasted 8 months and it focused on teaching the participants (6 elderly people) to design and program digital activities and materials, such as interactive stories, games or animations, through Scratch language.

6. Conclusions

Most of the participants in these works highlighted the numerous benefits and contributions of ICT in their daily lives, such as greater opportunities for leisure and communication, improvements in mood and social participation, stimulation of their abilities, mainly cognitive, improvements in their security and personal autonomy, or the possibility of making new learnings.

So, the results obtained in these projects show that technology can contribute to the improvement of quality of life of the elderly population, and to the strengthening of the four pillars of the AA concept: health, participation, safety and lifelong learning. However, it is necessary to develop further experiences and similar studies, in order to deepen the effects and opportunities offered by ICT to older people.

Author Contributions: L.N.-R. and T.P.G. conceived and designed the experiments; L.N.-R. performed the experiments; L.N.-R. and T.P.G. analyzed the data; T.P.G. and M.d.C.M.D. contributed materials and analysis tools; L.N.-R. and M.d.C.M.D. wrote the paper.

Acknowledgments: Part of this work was supported by: "In-TIC Project", promoted by Orange Foundation, with the support of the Plan Avanza (Spanish Ministry of Industry, Energy and Tourism), IMSERSO (Spanish Ministry of Health, Social Services and Equality), Fundació Agrupació Mútua, MAPFRE Foundation and La Caixa Foundation; "CloudPatient Project", co-funded by the Spanish Ministry of Economy and Competitiveness from FEDERINNTERCONECTA Program; "Geria-TIC Project" [IN852A 2016/10], co-funded by the Galician Innovation Agency (GAIN) from the Conecta PEME Program (3rd edition) with European Regional Development Funds (FEDER) of the European Union; "Collaborative Project in Genomic Data Integration (CICLOGEN)" [PI17/01826], funded by the Carlos III Health Institute from the Spanish National Plan for Scientific and Technical Research and Innovation 2013-2016 and the European Regional Development Funds (FEDER) "A way to build Europe" of the European Union; and aids for the consolidation and structuring of competitive research units [GRC2014/049, CN 2011/034] and singular centers [Centro singular de investigación de Galicia, accreditation 2016-2019, ED431G/01; CN 2012/211] of the Galician University System of the Xunta de Galicia. In addition, this work has received financial support from the Galician Research Network on Colorectal Cancer (REGICC) [ED431D 2017/23, R2014/039, CN 2012/217] and the Galician Network of Drugs R+D (REGID) [ED431D 2017/16], funded by the Xunta de Galicia.

Conflicts of Interest: The authors declare no conflict of interest. The founding sponsors had no role in the design of the study; in the collection, analyses, or interpretation of data; in the writing of the manuscript, and in the decision to publish the results.

References

1. World Health Organization (WHO). *Active Ageing: A Policy Framework*; WHO: Geneve, Switzerland, 2002; p. 12.
2. Fernández-Ballesteros, R.; Caprara, M.; Iñiguez, J.; García, L. Promoción del envejecimiento activo: Efectos del programa "Vivir con vitalidad"®. *Rev. Esp. Geriatr. Gerontol.* **2005**, *40*, 92–103.
3. Mendoza-Ruvalcaba, N.M.; Fernández-Ballesteros, R. Effectiveness of the vital aging program to promote active aging in Mexican older adults. *Clin. Interv. Aging* **2016**, *11*, 1631–1644.
4. Llorente-Barroso, C.; Viñarás-Abad, M.; Sánchez-Valle, M. Internet and the elderly: Enhancing active ageing. *Comunicar* **2015**, *23*, 29–36.
5. Riva, G.; Gaggioli, A.; Villani, D.; Cipresso, P.; Repetto, C.; Serino, S.; Triberti, S.; Brivio, E.; Galimberti, C.; Graffigna, G. Positive technology for healthy living and active ageing. *Stud. Health Technol. Inform.* **2014**, *203*, 44–56.
6. Riva, G.; Villani, D.; Cipresso, P.; Repetto, C.; Triberti, S.; Di Lernia, D.; Chirico, A.; Serino, S.; Gaggioli, A. Positive and transformative technologies for active ageing. *Stud. Health Technol. Inform.* **2016**, *220*, 308–315.
7. Nieto-Riveiro, L.; Groba, B.; Servia, F. Experiences using Information and Communication Technologies with elderly people. In *Handbook of Research on Personal Autonomy Technologies and Disability Informatics*; IGI Global: New York, NY, USA, 2011; pp. 346–357.

8. Groba, B.; Nieto-Riveiro, L. Integración de las tecnologías de la información y las comunicaciones. Experiencia en el centro de día de mayores de Cruz Roja de A Coruña. In *Aplicación de las Tecnologías de la Información y las Comunicaciones en la Vida Diaria de las Personas con Discapacidad*; Nieto-Riveiro, L., Groba, B., Pousada, T., Pereira, J., Eds.; Servizo de Publicacións da Universidade da Coruña, Orange Foundation: A Coruña, Spain, 2012; pp. 43–56.
9. Nieto-Riveiro, L. Study about the Impact of an Active Aging Program through Technological Tools. Ph.D. Thesis, Universidade da Coruña, A Coruña, Spain, 2016.

proceedings

MDPI

Extended Abstract

Laboratory Samples Allocation Problem †

Diego Noceda Davila *, Luisa Carpente Rodríguez and Silvia Lorenzo Freire

MODES Research Group, Department of Mathematics, University of A Coruña, A Coruña 15001, Spain; luisa.carpente@udc.es (L.C.R.); silvia.lorenzo@udc.es (S.L.F.)
* Correspondence: diego.noceda@udc.es; Tel.: +34-981-167-000
† Presented at the XoveTIC Congress, A Coruña, Spain, 27–28 September 2018.

Published: 14 September 2018

Abstract: This work aims to solve the optimization problem associated with the allocation of laboratory samples in plates. The processing of each of these plates is costly both in time and money, therefore the main objective is to minimize the number of plates used. The characteristics of the problem are reminiscent of the well-known bin packing problem, an NP-Hard problem that, although it is feasible to model as a linear programming problem, it cannot be solved at a reasonable cost. This work, proposes the implementation of a heuristic algorithm that provides good results at a low computational cost.

Keywords: optimization; planning problem; simulated annealing; linear programming; bin packing

1. Introduction

The daily work of a laboratory requires processing samples in thermal cyclers as part of more complex processes. As processing each of the plates is costly in both time and money, the main objective of the work is to reorganize the samples in the plates taking into account the restrictions imposed by the characteristics of the thermal cycler.

The plates used by the laboratory consist of 8 rows and 12 columns, making a total of 9 6 wells. In turn, the 12 columns are divided into 6 temperature bands with 16 wells each. The first restriction imposes that the difference of temperature between two adjacent strips must be less than 5 °C.

The input data of the problem has the following characteristics, each sample occupies a well, each group of samples is processed at a certain temperature and, in addition, for each group is necessary to reserve a well to accommodate the group control. Figure 1 shows an example plate of this configuration.

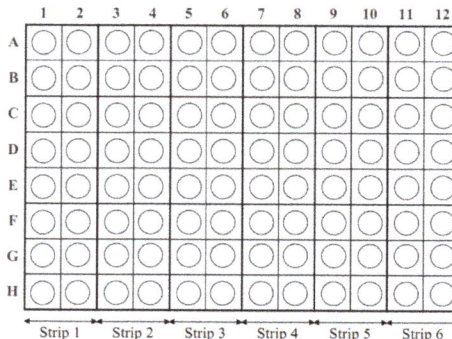

Figure 1. Example of laboratory plate.

The objective of the work is to obtain the distribution of the samples in plates in such a way that the number of plates used is minimized, the total number of cells used is minimized and the percentage of occupancy of the plates is maximized according to the lexicographical order.

2. The Algorithm

As seen in the article published by [1], since it is necessary to achieve a solution in a reasonable time, a heuristic algorithm, based on the simulated annealing method introduced by [2], has been developed. The idea of the method comes from the analogy between the process of metal annealing and the optimization of combinational problems. To implement this method, it is necessary to define the problem in terms of a solution space with a neighborhood and a solution space.

The algorithm starts from an initial solution, obtained by filling the plates with the samples ordered by temperature, bearing in mind that each group needs a present indicator, different groups with the same temperature can share a strip and it is possible to leave free strips to maintain the difference of 5 °C between consecutive strips. To this first solution, minimal changes are made (the neighbors), on which the objective function is calculated, which allows us to assess whether the change has been positive for the resolution of the problem. The movements that allow these changes are two, the exchange of strips and the grouping of samples dispersed by several plates.

3. Preliminary Results

The laboratory works with commercial software known as LabWare, against which the results of the implemented algorithm have been compared. Table 1 shows a summary of the solutions obtained for a set of tests, showing the number of plates necessary in each option, accompanied by the execution time of the heuristic algorithm

Table 1. Results and comparison.

Sample Count	LabWare	Heuristic Algorithm	Time Spent (s)
1473	22	19	71.25
1944	47	23	92.48
2071	41	25	59.03
2248	56	27	68.30
2496	36	30	83.08
2703	65	32	102.79
3783	90	44	109.73

4. New Challenges

In the real world, samples have different priorities (modeled by temporary windows), the processing capacity of plates is limited and sometimes it will be inevitable to postpone part of the work. In addition, the laboratory receives new samples to be processed periodically.

In order to face these new challenges, an architecture to generalize this work has been designed. This architecture must meet several requirements, such as maintaining the quality of the solutions, being able to carry out tasks in parallel with great autonomy and being of help to operators in making decisions.

This new approach consists of a web interface, capable of multiprocessing parallelism, scalable and able to work with the complete solution with which operators can interact. Figure 2 shows an example capture of the result of an execution.

Solution Detail:

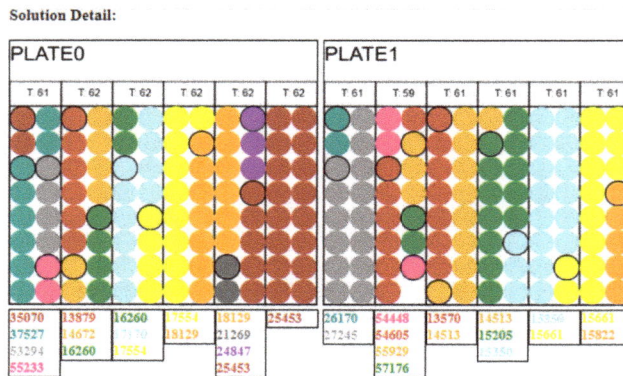

Figure 2. Example of algorithm output in the web interface.

Author Contributions: L.C.R. and S.L.F. conceived and designed the algorithm and the experiments. D.N.D. developed the new software architecture.

Acknowledgments: This work has been supported by MINECO grants MTM2014-53395-C3-1-P and MTM2017-87197-C3-1-P, and by Xunta de Galicia through the European Regional Development Fund-ERDF (Grupos de Referencia Competitiva ED431C-2016-015 and Centro Singular de Investigación de Galicia ED431G/01) and the European Social Fund-ESF.

Conflicts of Interest: The authors declare no conflict of interest. The founding sponsors had no role in the design of the study; in the collection, analyses, or interpretation of data; in the writing of the manuscript, and in the decision to publish the results.

References

1. Carpente, L.; Cerdeira-Pena, A.; Lorenzo-Freire, A.; Places, S. Optimization in Sanger sequencing. *Comput. Oper. Res.* under review.
2. Kirkpatrick, S.; Gelatt, C.D.; Vecchi, M.P. Optimization by simulated annealing. *Science* **1983**, *220*, 671–680.

proceedings

MDPI

Extended Abstracts

Sparse Semi-Functional Partial Linear Single-Index Regression †

Silvia Novo [1,*] , **Germán Aneiros** [1] and **Philippe Vieu** [2]

1   MODES Research Group, CITIC, Universidade da Coruña, 15071 A Coruña, Spain; german.aneiros@udc.es
2   Institut de Mathématiques, Université Paul Sabatier, 31062 Toulouse, France;
    philippe.vieu@math.univ-toulouse.fr
*   Correspondence: s.novo@udc.es; Tel.: +34-981-167-000
†   Presented at the XoveTIC Congress. A Coruña, Spain, 27–28 September 2018.

Published: 17 September 2018

check for updates

Abstract: The variable selection problem is studied in the sparse semi-functional partial linear model, with single-index type influence of the functional covariate in the response. The penalized least squares procedure is employed for this task. Some properties of the resultant estimators are derived: the existence (and rate of convergence) of a consistent estimator for the parameters in the linear part and an oracle property for the variable selection method. Finally, a real data application illustrates the good performance of our procedure.

Keywords: functional data analysis; variable selection; sparse model; dimension reduction; functional single-index model; semiparametric model

1. Introduction

In many real problems, to predict the value of a random variable, observations of many other variables are available. However, in many cases, it is unknown which of them (very few) have a real influence in the response. In this practical framework, we need procedures able to select the relevant variables to avoid high-dimensionality problems. Reducing the complexity of the model becomes even more crucial when regression involves a functional variable too (data are functions, images...). Therefore, the main goal is the simplification of the model, which makes easier both its estimation and interpretation, without losing its predictive efficiency.

These practical problems have motived the peak of semiparametric models in the functional regression, together with the variable selection procedures. In [1] the penalized least squares method for estimation and variable selection is studied for the partial linear model with functional covariate. In this model, the real variables have a linear effect (involving interpretable coefficients that are the parameters) in the response, while the infinite-dimensional covariate has a nonlinear (nonparametric) influence. However, in real data applications, it would be interesting having parameters related to the functional variable to derive practical interpretations. This is one of the advantages of the semi-functional partial linear single-index model (SFPLSIM): the real covariates also affect in a linear way to the response, but the infinite-dimensional covariate influences it trough a projection in an unknown direction, after applying a nonlinear link function. This direction of projection behaves like a function-parameter that could have interesting interpretations. Some theoretical properties related to the nonparametric estimation of the functional single-index model are given in [2]. In this paper, we will study the sparse SFPLSIM, focusing in the variable selection problem. For this purpose, we will use the penalized least squares procedure for estimating the parameters of the lineal components and, simultaneously, selecting the relevant covariates. The properties of the estimators will be analysed from a theoretical point of view: we will set its convergence rates and the consistency for selecting the model. These results will be illustrated through a real data application.

2. The Model

The SFPLSIM is defined by the relationship

$$Y_i = X_{i1}\beta_{01} + \cdots + X_{ip_n}\beta_{0p_n} + m\left(\langle \theta_0, \mathcal{X}_i \rangle\right) + \varepsilon_i, \ \forall i = 1, \ldots, n, \tag{1}$$

where Y_i denotes a scalar response, $X_{i1}, \ldots, X_{ip_n}$ are random covariates taking values in $\mathbb{R}$ and $\mathcal{X}_i$ is a functional random covariate valued in a separable Hilbert space $\mathcal{H}$ with inner product $\langle \cdot, \cdot \rangle$. $\boldsymbol{\beta}_0 = (\beta_{01}, \ldots, \beta_{0p_n})^\top \in \mathbb{R}^{p_n}$, $\theta_0 \in \mathcal{H}$ and $m(\cdot)$ are a vector of unknown real parameters, an unknown functional direction and an unknown smooth real-valued function, respectively. Finally, ε_i is the random error, which verifies $\mathbb{E}\left(\varepsilon_i | X_{i1}, \ldots, X_{ip_n}, \mathcal{X}_i\right) = 0$.

3. The Penalized Least-Squares Estimators

For the purpose of simultaneously estimating β-parameters and selecting relevant X-covariates in the SFPLSIM (1), we will apply the penalized least-squares approach. For that, in a first step we transform the SFPLSIM in a linear model by extracting from Y_i and X_{ij} ($j = 1, \ldots, p_n$) the effect of the functional covariate $\mathcal{X}_i$ when is projected on the direction θ_0. Specifically, denoting by $\boldsymbol{X}_i = \left(X_{i1}, X_{i2}, \ldots, X_{ip_n}\right)^\top$, $\boldsymbol{X} = (\boldsymbol{X}_1, \ldots, \boldsymbol{X}_n)^\top$ and $\boldsymbol{Y} = (Y_1, \ldots, Y_n)^\top$, the fact that

$$Y_i - \mathbb{E}\left(Y_i | \langle \theta_0, \mathcal{X}_i \rangle\right) = \left(\boldsymbol{X}_i - \mathbb{E}\left(\boldsymbol{X}_i | \langle \theta_0, \mathcal{X}_i \rangle\right)\right)^\top \boldsymbol{\beta}_0 + \varepsilon_i, \ \forall i = 1, \ldots, n, \tag{2}$$

allows to consider the following approximate linear model (see Appendix A for understanding the notation):

$$\widetilde{\boldsymbol{Y}}_{\theta_0} \approx \widetilde{\boldsymbol{X}}_{\theta_0} \boldsymbol{\beta}_0 + \boldsymbol{\varepsilon}, \tag{3}$$

where $\boldsymbol{\varepsilon} = (\varepsilon_1, \ldots, \varepsilon_n)^\top$. Then, in a second step, the penalized least-squares approach is applied to model (3). Specifically, $\boldsymbol{\beta}_0$ and θ_0 are estimated by considering a minimizer, $(\widehat{\boldsymbol{\beta}}_0, \widehat{\theta}_0)$, of the penalized profile least-squares function

$$\mathcal{Q}(\boldsymbol{\beta}, \theta) = \frac{1}{2}\left(\widetilde{\boldsymbol{Y}}_\theta - \widetilde{\boldsymbol{X}}_\theta \boldsymbol{\beta}\right)^\top \left(\widetilde{\boldsymbol{Y}}_\theta - \widetilde{\boldsymbol{X}}_\theta \boldsymbol{\beta}\right) + n \sum_{j=1}^{p_n} \mathcal{P}_{\lambda_{j_n}}\left(|\beta_j|\right),$$

where $\boldsymbol{\beta} = (\beta_1, \ldots, \beta_{p_n})^\top$, $\mathcal{P}_{\lambda_{j_n}}(\cdot)$ is a penalty function and $\lambda_{j_n} > 0$ is a tuning parameter. Note that, simultaneously to the parameter estimation, the previous procedure can be considered as a variable selection method: if $\widehat{\beta}_{0j}$ is a non-null component of $\widehat{\boldsymbol{\beta}}_0$, then X_j is selected as an influential variable.

From now on, we will denote $J_n = \{1, \ldots, p_n\}$ and $S_n \subset J_n$ such that $\beta_{0j} \neq 0$ for $j \in S_n$ and $\beta_{0j} = 0$ for $j \in S_n^c = J_n / S_n$. In addition s_n will mean $\operatorname{card}(S_n)$ and we will assume that $S_n = \{1, \ldots, s_n\}$.

4. Asymptotic Theory

In this paper, the existence of the penalized estimator is established as well as the corresponding rates of convergence. In particular, under some assumptions, we proved that there exists a local minimizer $(\widehat{\boldsymbol{\beta}}_0, \widehat{\theta}_0)$ of $\mathcal{Q}(\boldsymbol{\beta}, \theta)$ such that

$$\left\|\widehat{\boldsymbol{\beta}}_0 - \boldsymbol{\beta}_0\right\| = O_p\left(\sqrt{s_n}\left(n^{-1/2} + \delta_n\right)\right) \text{ where } \delta_n = \max_{j \in S_n}\left\{\left|\mathcal{P}'_{\lambda_{j_n}}\left(|\beta_{0j}|\right)\right|\right\}. \tag{4}$$

Furthermore, the selected set of variables, $\widehat{S}_n = \{j \in J_n; \widehat{\beta}_{0j} \neq 0\}$, works as well (at least asymptotically) as it would do if the true set of relevant variables S_n was known. Specifically, $\mathbb{P}(\widehat{S}_n = S_n) \to 1$ as $n \to \infty$.

An application to real data is included, which shows the good performance of the presented method in terms of error of prediction.

Funding: The authors acknowledge partial support by MINECO grants MTM2014-52876-R and MTM2017-82724-R (EU ERDF support included). Additionally, financial support from the Xunta de Galicia (Centro Singular de Investigación de Galicia accreditation ED431G/01 2016-2019 and Grupos de Referencia Competitiva ED431C2016-015) and the European Union (European Regional Development Fund - ERDF), is gratefully acknowledged. The first author also thanks the financial support from the Xunta de Galicia and the European Union (European Social Fund - ESF), the reference of which is ED481A-2018/191.

Conflicts of Interest: The authors declare no conflict of interest. The founding sponsors had no role in the design of the study; in the collection, analyses, or interpretation of data; in the writing of the manuscript, and in the decision to publish the results.

Abbreviations

The following abbreviations are used in this manuscript:

SFPLSIM   Semi-functional partial linear single index model

Appendix A. Notation

For any $(n \times q)$-matrix $\boldsymbol{A}$ $(q \geq 1)$, if $\boldsymbol{I}$ is the $(n \times n)$-identity-matrix, we denote

$$\widetilde{\boldsymbol{A}}_\theta = (\boldsymbol{I} - \boldsymbol{W}_{h,\theta}) \, \boldsymbol{A}, \text{ where } \boldsymbol{W}_{h,\theta} = \left(w_{n,h,\theta}(\mathcal{X}_i, \mathcal{X}_j)\right)_{i,j},$$

with $w_{n,h,\theta}(\cdot, \cdot)$ being the weight function

$$w_{n,h,\theta}(\chi, \mathcal{X}_i) = \frac{K\left(d_\theta\left(\chi, \mathcal{X}_i\right)/h\right)}{\sum_{j=1}^n K\left(d_\theta\left(\chi, \mathcal{X}_j\right)/h\right)},$$

where $K : \mathbb{R}^+ \to \mathbb{R}^+$ is a kernel function, $h > 0$ is a smoothing parameter and, for $\theta \in \mathcal{H}$, $d_\theta(\cdot, \cdot)$ is the semimetric defined as

$$d_\theta\left(\chi, \chi'\right) = \left|\langle\theta, \chi - \chi'\rangle\right|, \; \forall \chi, \chi' \in \mathcal{H}.$$

References

1. Aneiros, G.; Ferraty, F.; Vieu, P. Variable selection in partial linear regression with functional covariate. *Statistics* **2015**, *49*, 1322–1347, doi:10.1080/02331888.2014.998675.
2. Novo, S.; Aneiros, G.; Vieu, P. Automatic and location-adaptive estimation in functional single-index regression. **2018**, in press.

proceedings

MDPI

Extended Abstract

Comparative Results with Unsupervised Techniques in Cyber Attack Novelty Detection †

Jorge Meira

Computing Department, University of Coruña, Coruña 15071, Spain; j.a.meira@udc.es; Tel.: +34-981-167-000
† Presented at the XoveTIC Congress, A Coruña, Spain, 27–28 September 2018.

Published: 17 September 2018

Abstract: Intrusion detection is a major necessity in current times. Computer systems are constantly being victims of malicious attacks. These attacks keep on exploring new technics that are undetected by current Intrusion Detection Systems (IDS), because most IDS focus on detecting signatures of previously known attacks. This work explores some unsupervised learning algorithms that have the potential of identifying previously unknown attacks, by performing outlier detection. The algorithms explored are one class based: the Autoencoder Neural Network, K-Means, Nearest Neighbor and Isolation Forest. There algorithms were used to analyze two publicly available datasets, the NSL-KDD and ISCX, and compare the results obtained from each algorithm to perceive their performance in novelty detection.

Keywords: unsupervised learning; anomaly detection; outlier detection; novelty detection

---

1. Introduction

Cyber-Security is a field that is constantly evolving, the rate by which new threats and attacks appear is enormous and this requires a constant research for vulnerabilities and ways of solving them by the people responsible for the Security Systems [1].

Intrusion Detection Systems (IDS) are tools based on attack detection techniques to finding out new vulnerabilities. IDS tend to follow one of two different approaches: signature based, or anomaly based. Signature based detection requires previous knowledge of an attack before being able to identify it, on the other hand anomaly base detection only requires knowledge of regular data, and any potential deviation from that norm can correspond to an attack, even if the attack has not been discovered yet [2]. This is an arduous task, and classification algorithms can be used to aid in this scenario. Some algorithms, called supervised learning algorithms are well suited for problems where exiting classified examples can be used as training data for the algorithm. However, with new vulnerabilities there are no classified examples for supervised algorithms to learn from. One possibility to help with this problem is the usage of unsupervised learning algorithms. Unsupervised learning algorithms can learn what is normal data and find deviations from that, which in this case would indicate a possible attack previously unknown (anomaly based).

2. Experimental Work

In this work, was studied the behavior of some unsupervised algorithms based in one class classification, to verify if these techniques are a viable solution to discover and detect unknown attacks. In this section is presented and described the network anomaly detection workflow as shown in Figure 1.

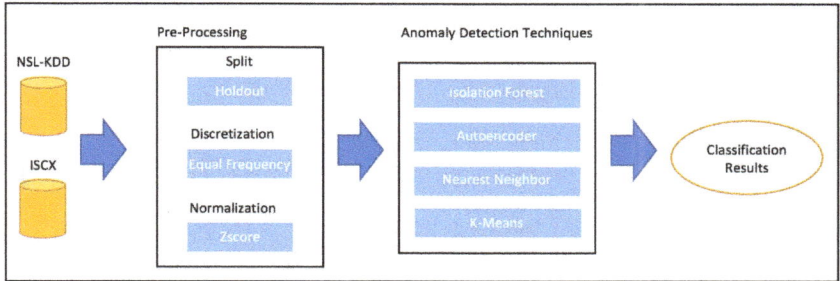

Figure 1. Anomaly detection workflow.

Datasets and Preprocessing

In our exploration, we analyzed the NSL-KDD [3,4] and the ISCX datasets [5]. These datasets contain samples from normal activity and from simulated attacks in computer systems and are commonly used in the literature. Before using the learning algorithms, we've applied some pre-processing methods to prepare the data. As shown in Figure 1, it was first applied the holdout method to both datasets, where 2/3 of the data (corresponding to normal activity in network) were used to train the algorithms and 1/3 of the data where 10% of this portion corresponds to anomalies, were used to test the algorithms. The next step was discretization, where all continuous features of both datasets were converted to categorical features trough the equal frequency technique. The last pre-processing method applied was the data normalization, to have all the features within the same scale. This way, prevents some classification algorithms to give more importance at features with large numeric values. Z-Score was the normalization technique applied to the data. This technique transforms the input, so the mean is zero and the standard deviation is one. After the data cleaning and transformation, we applied four one-class algorithms, namely Autoencoder, Nearest Neighbor, K-Means and Isolation Forest and evaluate is performance in the NSL-KDD and ISCX datasets.

3. Comparative Evaluation and Conclusions

We tested all combinations of pre-processing techniques with the unsupervised learning algorithms and graphically presented the results of the best techniques applied to each algorithm for NSL-KDD and ISCX datasets.

In both datasets all the algorithms had a high accuracy. That was expected as most of the samples are from normal activity. To achieve better conclusions about the algorithms efficacy the F1 metric will also be analyzed. Starting by the NSL-KDD dataset shown in Figure 2a, we can see that all of the algorithms had a similar result close to 60% of F1. It is not a high-performance score, however the recall was much higher than precision in K-Means, Nearest Neighbor and Isolation forest algorithms, around 80%, which means that the false negatives were much less than the false positives. In cybersecurity it is important to have a low false negative rate, since it represents data predicted as normal, while in fact it represents malicious or abnormal activity.

In the ISCX dataset (Figure 2b) all algorithms showed a slightly better performance, except the Nearest Neighbor algorithm which had a much higher score compared to the NSL-KDD. These results were expected in the ISCX, whereas this dataset only has 4 different types of attacks compared to the 38 different types in the NSL-KDD dataset.

The results showed that these unsupervised techniques combined with the best preprocessing techniques could detect most of the anomaly instances but also generate a lot of false positives. These occur due to the similarity between normal and abnormal instances.

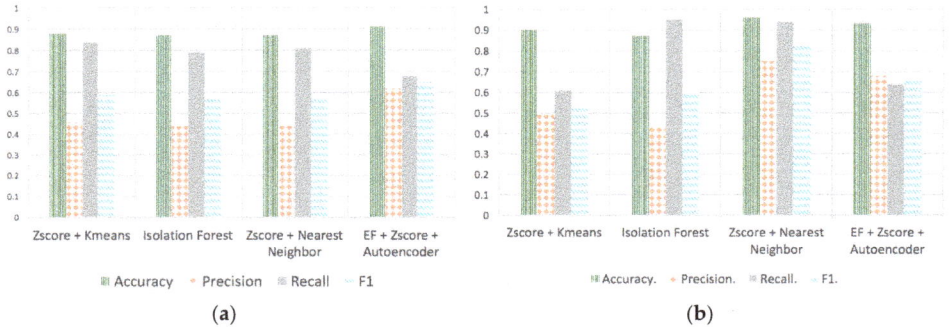

Figure 2. Anomaly detection results in (**a**) NSL-KDD and (**b**) ISCX datasets.

Author Contributions: His work was conceived and performed by J.M. and guided by Amparo Alonso Betanzos, Veronica Bolón Canedo and Goreti Marreiros, the colleagues from the research group GECAD and LIDIA helped in the reviewing process of this work.

Acknowledgments: This is an extension of the work made in the SASSI project (ANI | P2020 17775). This work is supported by the research project ED431G/01, and has received funding from the Xunta de Galicia (Centro singular de investigación de Galicia accreditation 2016-2019) and the European Union (European Regional Development Fund-ERDF).

Conflicts of Interest: The authors declare no conflict of interest. The founding sponsors had no role in the design of the study; in the collection, analyses, or interpretation of data; in the writing of the manuscript, and in the decision to publish the results.

References

1. Zanero, S.; Serazzi, G. Unsupervised learning algorithms for intrusion detection. In Proceedings of the 2008 IEEE Network Operations and Management Symposium, Salvador, Bahia, Brazil, 7–11 April 2008; pp. 1043–1048.
2. Casas, P.; Mazel, J.; Owezarski, P. Unsupervised Network Intrusion Detection Systems: Detecting the Unknown without Knowledge. *Comput. Commun.* **2012**, *35*, 772–783.
3. Noto, K.; Brodley, C.; Slonim, D. FRaC: A feature-modeling approach for semi-supervised and unsupervised anomaly detection. *Data Min. Knowl. Discov.* **2012**, *25*, 109–133.
4. *KDD Cup 1999 Data*; UCI Machine Learning Repository: Irvine, CA, USA, 1999. Available online: http://archive.ics.uci.edu/ml/datasets/kdd+cup+1999+data (accessed on 1 January 2018).
5. Shiravi, A.; Shiravi, H.; Tavallaee, M.; Ghorbani, A.A. Toward developing a systematic approach to generate benchmark datasets for intrusion detection. *Comput. Secur.* **2012**, *31*, 357–374.

proceedings

Extended Abstract

A Vehicle Routing Problem with Periodic Replanning †

Guido Ignacio Novoa-Flores *, Luisa Carpente and Silvia Lorenzo-Freire

MODES Research Group, Department of Mathematics, Faculty of Computer Science, University of A Coruña, 15071 A Coruña, Spain; luisacar@udc.es (L.C.); slorenzo@udc.es (S.L.-F.)
* Correspondence: guido.novoa@udc.es; Tel.: +34-981-167-000
† Presented at the XoveTIC Congress, A Coruña, Spain, 27–28 September 2018.

Published: 19 September 2018

Abstract: In this work we focus on the problem of truck fleet management of the company GESUGA. This company is responsible of the collection and proper treatment of animals not intended for human consumption. On a daily basis, with the uncollected requests, the company designs the routes for the next day. However, these routes have to be replanned during their execution as new requests appear from customers that the company would be interested in attending. The problem treated belongs to the family MDCVRPTW with the particularity of the route redesign. For its resolution we have adapted linear programming models, simulation techniques and metaheuristic algorithms.

Keywords: combinatorial optimization; heuristic algorithms; vehicle routing problems

1. Introduction

After the spread of Bovine Spongiform Encephalopathy, typically known as mad cow disease, the European Union took a number of measures (e.g., Regulation (EC) No 999/2001) to prevent its spread and transmission. This regulation forbids the burial of carcasses of dead animals at livestock farms. In Galicia, one of the main companies responsible for this task is GESUGA. This company focuses its business area on the integrated management of meat by-products not intended for human consumption. Its main activity consists of the collection and transport of the different meat by-products, generally animal carcasses, from livestock farms to treatment plants, for their appropriate treatment.

To serve customers, the fleet of the company is composed by 32 trucks (12 in Cerceda, 10 in Outeiro de Rei and 10 in Vilamarín) which, from Monday to Friday, visit the different farms and transport the products to the intermediate plants. Taking into account the characteristics of this problem, it could be classified as a MDCVRPTW. A general review of VRP can be found in [1].

2. Description of the Problem

As mentioned in the introduction, the company has to visit its customers all over Galicia on a daily basis. Some of the restrictions that define this problem are the following:

- The trucks leave and return from the plant to which they are assigned only once a day.
- Truck drivers have a maximum working day of 8 h which includes a rest and disinfection of the vehicle at the end of the day.
- Trucks have a maximum loading capacity.
- Orders must be picked up within 48 h from receipt.

Currently, route planning is manually made by the logistics department and the organization is as follows:

- At 19:00 there are some pre-routes with the notices not collected until that moment.
- At 20:00 these pre-routes are reviewed with the logistics manager adding new requests and making the necessary changes.
- At 21:00 drivers receive the set of places that they must visit, but they are free to organize it.
- During the day, incoming requests are assigned manually by the logistics department to drivers in order to free up work for the next day.

Note that the route design is manually made by the logistics department. Therefore, the company is interested in a tool to calculate the routes automatically, satisfy the needs of customers and achieve the following objectives:

- Minimize the total distance traveled by trucks.
- Minimize the number of trucks used.
- Maximize the number of collected requests.

3. Implementation of the Algorithm

The implementation of the algorithm was made in JAVA language using the libraries lpsolve and jsprit. The second library includes the Ruin and Recreate principle (see [2] and strategies inspired by [3]. The problems mentioned above are solved automatically according to the following scheme:

1. Requests that are not collected during a day are assigned to a plant by solving the GAP problem with lpsolve library.
2. For each plant, the corresponding VRPs are resolved with jsprit library.
3. The requests that arrive online are assigned to each truck automatically taking into account the position and the load of each truck.

Currently, we are considering two strategies to address this problem:

- Lazy: No orders are collected during the online phase, i.e., the routes computed the day before collecting are not modified.
- Minimum-k: A truck leaves the plant when, at least, k orders are assigned to it.

Note that the Lazy strategy can only be used from Monday to Thursday since on Fridays no orders can be left uncollected. Thus, Lazy strategy must be combined with Minimum-k.

4. Results

Many scenarios have been considered for the different strategies varying different parameters: cost of taking out a truck, minimum number of orders needed to take out a truck and time at which the optimization online is performed.

The appendix shows the best results obtained for each of the strategies as well as the real case. We see that the real case (Table A1) collects 6040 requests using 309 trucks. The Lazy strategy (Table A2) can collect 5782 requests using 275 trucks and the Minimum-k (Table A3) collects 6224 using 304 trucks. Therefore, we can conclude that the Lazy strategy always picks up fewer requests and uses fewer trucks than the real case. On the other hand, the Minimum-k strategy collects more requests and uses more trucks than the real case but the proportion between trucks and collected requests improves with respect to the real case.

Author Contributions: The first author carried out the experiments and the implementation of the algorithm. The remaining authors were responsible for the design of the algorithm.

Acknowledgments: This work has been supported by MINECO grants MTM2014-53395-C3-1-P and MTM2017-87197-C3-1-P, by the Centre for the Development of Industrial Technology through the proyect ITC-20151247 and by Xunta de Galicia through the European Regional Develpoment Fund-ERDF (Grupos de

Referencia Competitiva ED431C-2016-015 and Centro Singular de Investigaci\'on de Galicia ED431G/01) and the European Social Fund-ESF.

Conflicts of Interest: The authors declare no conflict of interest. The founding sponsors had no role in the design of the study; in the collection, analyses, or interpretation of data; in the writing of the manuscript, and in the decision to publish the results.

Appendix A

The following abbreviations are used in this manuscript:

- GESUGA Gestora de Subproductos de Galicia.
- MDCVRPTW Multi Depot Capacitated Vehicle Routing Problem with Time Windows
- VRP Vehicle Routing Problem

Appendix B

Table A1. Results obtained by the company.

Day	Requests	Previous Requests	Online	Collected Requests	Trucks
31 August 2016	675	271	404	527	29
01 September 2016	624	148	476	482	28
02 September 2016	563	142	421	504	29
05 September 2016	1126	362	764	639	28
06 September 2016	950	487	463	644	28
07 September 2016	757	306	451	569	31
08 September 2016	628	187	441	480	27
09 September 2016	584	148	436	529	29
12 September 2016	1011	313	698	567	27
13 September 2016	908	444	464	564	27
14 September 2016	754	344	410	535	26
Total	8580	3152	5428	6040	309

Table A2. Results obtained with Lazy strategy.

Day	Requests	Previous Requests	Online	Collected Requests	Trucks
31 August 2016	675	271	404	271	17
01 September 2016	880	404	476	404	20
02 September 2016	897	476	421	706	32
05 September 2016	1267	503	764	503	23
06 September 2016	1227	764	463	725	31
07 September 2016	953	502	451	502	24
08 September 2016	892	451	441	451	21
09 September 2016	877	441	436	702	32
12 September 2016	1144	446	698	446	21
13 September 2016	1162	698	464	685	31
14 September 2016	887	477	410	477	23
Total	10861	5433	5428	5872	275

Table A3. Results obtained with Minimum-k strategy.

Day	Requests	Previous Requests	Online	Collected Requests	Trucks
31 August 2016	614	138	476	403	24
01 September 2016	632	211	421	465	25
02 September 2016	1243	479	764	760	32
05 September 2016	946	483	463	702	32
06 September 2016	695	244	451	536	30
07 September 2016	600	159	441	450	25
08 September 2016	586	150	436	426	24
09 September 2016	1129	431	698	733	32
12 September 2016	860	396	464	653	31
13 September 2016	617	207	410	459	27
14 September 2016	614	138	476	403	24
Total	8597	3169	5428	6124	310

References

1. Toth, P.; Vigo, D. *Vehicle Routing: Problems, Methods, and Applications,* 2nd. ed.; SIAM: Philadelphia, PA, USA, 2014.
2. Schrimpf, G.; Schneider, J.; Stamm-Wilbrandt, H.; Dueck, G. Record Breaking Optimization Results Using the Ruin and Recreate. *J. Comput. Phys.* **2000**, *159*, 139–171.
3. Pisinger, D.; Ropke S. A general heuristic for vehicle routing problems. *Comput. Oper. Res.* **2007**, *34*, 2403–2435.

proceedings

MDPI

Extended Abstract

Scene Wireframes Sketching for Drones [†]

Roi Santos *, Xose M. Pardo and Xose R. Fdez-Vidal

CITIUS, Rúa de Jenaro de la Fuente, s/n, Santiago de Compostela, 15705 A Coruña, Spain;
xose.pardo@usc.es (X.M.P.); xose.vidal@usc.es (X.R.F-V.)
* Correspondence: roi.santos@usc.es; Tel.: +34-8818-16443
† Presented at the XoveTIC Congress, A Coruña, Spain, 27–28 September 2018.

Published: date

Abstract: The increasing use of autonomous UAVs inside buildings and around human-made structures demands new accurate and comprehensive representation of their operation environments. Most of the 3D scene abstraction methods use invariant feature point matching, nevertheless some sparse 3D point clouds do not concisely represent the structure of the environment. Likewise, line clouds constructed by short and redundant segments with inaccurate directions limit the understanding of scenes as those that include environments with poor texture, or whose texture resembles a repetitive pattern. The presented approach is based on observation and representation models using the straight line segments, whose resemble the limits of an urban indoor or outdoor environment. The goal of the work is to get a full method based on the matching of lines that provides a complementary approach to state-of-the-art methods when facing 3D scene representation of poor texture environments for future autonomous UAV.

Keywords: 3D abstraction; reconstruction; line-based sketching; UAV

1. Introduction

The vast majority of the current approaches for 3D scene reconstruction are based on point clouds. Commonly, points are matched between pairs of views based on their descriptors, then triangulated [1] to make an initial estimation of their location in 3D space, and finally their poses are adjusted by least squares minimization [2]. A number of efficient point detectors and descriptor have made it possible to obtain robust and detailed 3D reconstructions based on feature point clouds [3–6]. These algorithms made possible to evolve from simple 3D reconstructions of the surfaces [7] to dense point reconstructions of extensive landscapes and cities [8].

The goal of this work is to obtain a real-time three dimensional representation of a scene by using a limited number of matched straight segments. Our approach takes advantage of multi-scale line detection and matching [9] to increase the accuracy of the line endpoints triangulation among pairs of line-matched frames. Secondly, our method goes one step ahead in the least squares adjustment of cameras and lines by exploiting geometrical relationships of the coplanar lines. After classifying the spatial lines according to their co-planarity, the intersection of the observed lines are brought into a second run of the SBA process.

2. Materials and Methods

The first problem to solve for the computation of a 3D sketch from the matched observations is that the camera poses P are unknown. These can be estimated from the endpoint correspondences of l of from a feature points based SfM pipeline. The first camera is provided with the pose $P^1 = K \times [I|0]$, being K the calibration matrix. The rest of cameras will be stacked from this position in the world reference frame. Once we have the camera matrices for the first pair of cameras, a linear triangulation method [10] can be used to retrieve the first estimations for 3D lines, i.e., the members of Γ with observed counterpart on both camera planes. The final spatial segments are obtained as the centre of gravity of their estimations obtained in the stereo triangulations. Finally, a Bundle Adjustment is performed to optimize the relative pose for the cameras and spatial lines. The flow for the proposed method is depicted in Figure 1.

3. Results

Figure 1 shows the result of the proposed method employing 8 images from the public Ground Truth dataset [11]. It is compared compare to the results of Line3D++ [12], shown in Figure 2. The result proves that the proposed method is able to obtain a number of structures of the house from a low number of images, and still holding a decent accuracy. On the other hand, the method Line3D++ [12] returns sparse short segments. This sparsity complicates the understanding of what the spatial line cloud is resembling, and difficult the alignment to the Ground Truth mesh. Note that this method also fails to retrieve any long segment of the house for this test case.

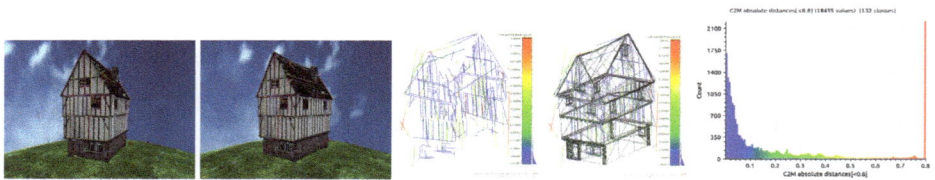

Figure 1. Resulting line matching using the proposed method for images {5,8} of the dataset [11]. Resulting 3D abstraction and measurements of distances to Ground Truth mesh.

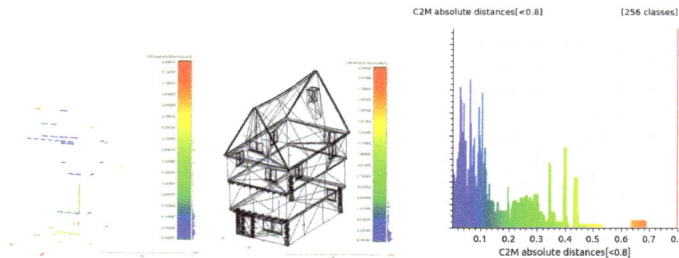

Figure 2. Results with the method Line3D++[12], using the same set of images as input.

4. Conclusions

This work presents a novel integration of a set of algorithms to create a line-based spatial sketch, showing the main structures of the man-made environment laying in front of a camera. It gets as input its intrinsic parameters and at least 3 pictures. The set of methods include novel observation relations of groups of straight segments that are captured from different poses. Quantitative results have been obtained and compared with other state-of-the-art line based SfM method. Future work might include the exploitation of weak epipolar constraints during the line matching process.

Author Contributions: Conceptualization, R.S., X.M.P. and X.R.F.-V.; Methodology, R.S., X.M.P. and X.R.F.-V.; Software, R.S.; Validation, R.S., X.M.P. and X.R.F.-V.; Formal Analysis, R.S., X.M.P. and X.R.F.-V. Writing—Review & Editing, R.S., X.M.P. and X.R.F.-V.; Supervision, X.M.P. and X.R.F.-V.

Funding: This work has received financial support from the Xunta de Galicia (Centro singular de investigación de Galicia accreditation 2016–2019) and the European Union (European Regional Development Fund–ERDF).

Acknowledgments: This work has received financial support from the Xunta de Galicia (Centro singular de investigación de Galicia accreditation 2016–2019) and the European Union (European Regional Development Fund–ERDF).

Conflicts of Interest: The authors declare no conflict of interest. The founding sponsors had no role in the design of the study; in the collection, analyses, or interpretation of data; in the writing of the manuscript, and in the decision to publish the results.

Abbreviations

The following abbreviations are used in this manuscript:

UAV   Unmanned Aereal Vehicle
SfM   Structure-From-Motion

References

1. Hartley, R.I.; Sturm, P. Triangulation. *Comput. Vis. Image Underst.* **1997**, *68*, 146–157.
2. Triggs, B.; Mclauchlan, P.; Hartley, R.; Fitzgibbon, A. Bundle Adjustment—A Modern Synthesis. In Proceedings of the ICCV '99 International Workshop on Vision Algorithms: Theory and Practice, Corfu, Greece, 21–22 September 1999; Volume 1, pp. 298–372.
3. Lowe, D.G. Distinctive Image Features from Scale-Invariant Keypoints. *Int. J. Comput. Vis.* **2004**, *2*, 91–110.
4. Bay, H.; Ess, A.; Tuytelaars, T.; Van Gool, L. Speeded-up robust features (SURF). *Comput. Vis. Image Underst.* **2008**, *110*, 346–359.
5. Rublee, E.; Rabaud, V.; Konolige, K.; Bradski, G. ORB: An efficient alternative to SIFT or SURF. In Proceedings of the 2011 IEEE International Conference on IEEE, Computer Vision (ICCV), Kona, HI, USA, 5–7 January 2011; pp. 2564–2571.
6. Alcantarilla, P.; Bartoli, A.; Davison, A. KAZE features. In Proceedings of the Computer Vision–ECCV 2012, Florence, Italy, 7–13 October 2012; pp. 214–227.
7. Pollefeys, M.; Van Gool, L.; Vergauwen, M.; Verbiest, F.; Cornelis, K.; Tops, J.; Koch, R. Visual modeling with a hand-held camera. *Int. J. Comput. Vis.* **2004**, *59*, 207–232.
8. Snavely, N.; Seitz, S.M.; Szeliski, R. Photo tourism: Exploring photo collections in 3D. *ACM Trans. Graph. (TOG)* **2006**, *25*, 835–846.
9. López, J.; Santos, R.; Fdez-Vidal, X.R.; Pardo, X.M. Two-view line matching algorithm based on context and appearance in low-textured images. *Pattern Recognit.* **2015**, *48*, 2164–2184.
10. Hartley, R.; Zisserman, A. *Multiple View Geometry in Computer Vision*, 2nd ed., Cambridge Press, 2004.
11. Jain, A.; Kurz, C.; Thormählen, T.; Seidel, H.P. Exploiting Global Connectivity Constraints for Reconstruction of 3D Line Segment from Images. In Proceedings of the IEEE Conference on Computer Vision and Pattern Recognition (CVPR 2010), San Francisco, CA, USA, 13–18 June 2010.
12. Hofer, M.; Maurer, M.; Bischof, H. Efficient 3D scene abstraction using line segments. *Comput. Vis. Image Underst.* **2016**, *157*, 167–178.

proceedings

MDPI

Extended Abstract

Automatic Identification and Segmentation of Diffuse Retinal Thickening Macular Edemas Using OCT Imaging [†]

Gabriela Samagaio [1,2], Joaquim de Moura [1,2,*], Jorge Novo [1,2] and Marcos Ortega [1,2]

[1] Department of Computing, University of A Coruña, 15071 A Coruña, Spain;
gabriela.samagaio@udc.es (G.S.); jnovo@udc.es (J.N.); mortega@udc.es (M.O.)

[2] CITIC-Research Center of Information and Communication Technologies, University of A Coruña,
15071 A Coruña, Spain

* Correspondence: joaquim.demoura@udc.es; Tel.: +34-981167000

† Presented at the XoveTIC Congress, A Coruña, Spain, 27–28 September 2018.

Published: 18 September 2018

Abstract: This paper proposes a novel methodology for the automatic identification and segmentation of the Diffuse Retinal Thickening (DRT) edemas using Optical Coherence Tomography (OCT) images as source of information. This Macular Edema (ME) type is commonly used by ophthalmologists as a relevant biomarker for the early diagnosis of this retinal disorder which, therefore, permits a better adjustment of the treatments, reducing their costs as well as improving the life quality of the patients.

Keywords: computer-aided diagnosis; Optical Coherence Tomography; Diffuse Retinal Thickening region; segmentation

1. Introduction

Diabetic retinopathy is one of the leading causes of vision impairment that affects 1% of the world population [1]. Diffuse Retinal Thickening (DRT) is a Macular Edema (ME) type derived from the local intraretinal fluid accumulation in the lower retinal layers. As illustrated in Figure 1, the presence of this edema produces profound structural and morphological modifications in the eye fundus, as an increment of the lower retinal layers. The absence of a limiting membrane allows the fluid to spread over the outer retinal region, leading to a single and continuous region.

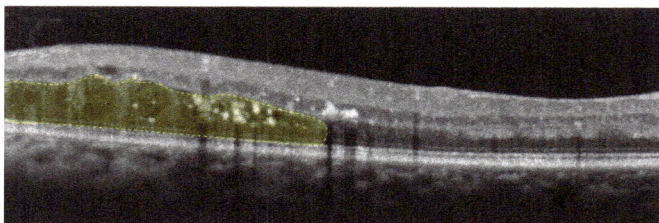

Figure 1. Example of OCT image with the DRT edema presence.

To identify the presence of retinal disorders, Optical Coherence Tomography (OCT) imaging is being widely used within the ophthalmological community. Moreover, it offers an easy visualization of the in vivo histopathology of the retina in a contactless and non-invasive capture process.

In this paper, we propose a new methodology for the automatic identification and segmentation of the DRT presence in OCT images using as reference the clinical classification [2,3]. The precise localization and delimitation of the DRT allow the early diagnosis of this disease and consequently, it permits a better adjustment of the treatments, reducing their costs as well as improving the life quality of the patients.

2. Methodology

The proposed methodology is composed by three main stages for the identification and segmentation of DRT edemas [4]. Firstly, the system segments the retinal layers to facilitate the search of this ME type in the Region of Interest (ROI) where they typically appear, the outer retina. Secondly, a learning strategy is applied to identify the DRT presence and segment its constituent region. Finally, two post-processing strategies are implemented and tested to individually refine the impact of the False Positives (FPs) and the False Negatives (FNs) detected regions from the classifier output and improve the obtained results.

3. Results and Conclusions

The proposed methodology achieved satisfactory results for the automatic identification and segmentation of the DRT edemas. Using the best classifier configuration, we applied two individuals post-processing strategies to improve the obtained results. Figure 2 illustrates the resulting image provided by the second post-processing, as the best configuration. This strategy unifies non-consecutive DRT regions, improving significantly the efficiency of the system and consequently facilitating the identification and visualization of the area affected by this pathology.

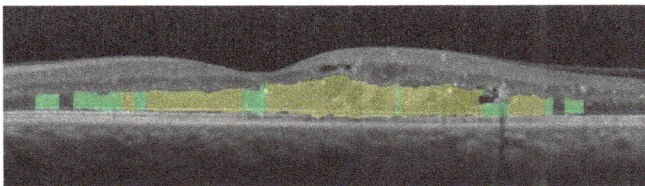

Figure 2. Illustrative output OCT image after the application of the second post-processing approach. Yellow regions, direct results from the classifier. Green regions, results of the second post-processing approach.

Author Contributions: G.S., J.d.M. and J.N. contributed to the development of the designed methodology and the analysis of the experimental evaluation methods. M.O. supported with the domain-specific knowledge. All the authors supervised and analyzed the obtained results. G.S. was responsible for writing the manuscript. All the authors cooperated in the critical revision and final approval of the manuscript structure.

Acknowledgments: This work is supported by the Instituto de Salud Carlos III, Government of Spain and FEDER funds of the European Union through the PI14/02161 and the DTS15/00153 research projects and by the Ministerio de Economía y Competitividad, Government of Spain through the DPI2015-69948-R research project. Also, this work has received financial support from the European Union (European Regional Development Fund-ERDF) and the Xunta de Galicia, Centro singular de investigación de Galicia accreditation 2016-2019, Ref. ED431G/01; and Grupos de Referencia Competitiva, Ref. ED431C 2016-047.

Conflicts of Interest: The authors declare no conflict of interest. The founding sponsors had no role in the design of the study; in the collection, analyses, or interpretation of data; in the writing of the manuscript, and in the decision to publish the results

References

1. World Health Organization. *World Health Statistics 2010*; World Health Organization: Geneva, Switzerland, 2010.
2. Otani, T.; Kishi, S.; Maruyama, Y. Patterns of diabetic macular edema with optical coherence tomography. *Am. J. Ophthalmol.* **1999**, *127*, 688–693.
3. Panozzo, G.; Parolini, B.; Gusson, E.; Mercanti, A.; Pinackatt, S.; Bertoldo, G.; Pignatto, S. Diabetic macular edema: An OCT-based classification. *Semin. Ophthalmol.* **2004**, *19*, 13–20.
4. Samagaio, G.; de Moura, J.; Novo, J.; Ortega, M. Automatic segmentation of diffuse retinal thickening edemas using Optical Coherence Tomography images. *Procedia Comput. Sci.* **2018**, *126*, 472–481.

proceedings

MDPI

Extended Abstract

Learning Retinal Patterns from Multimodal Images †

Álvaro S. Hervella 1,2,*, **José Rouco** 1,2, **Jorge Novo** 1,2 **and Marcos Ortega** 1,2

1 CITIC—Research Center of Information and Communication Technologies, University of A Coruña, 17051 A Coruña, Spain; jrouco@udc.es (J.R.); jnovo@udc.es (J.N.); mortega@udc.es (M.O.)
2 Department of Computer Science, University of A Coruña, 17051 A Coruña, Spain
* Correspondence: a.suarezh@udc.es; Tel.: +34-981-167-000
† Presented at the XoveTIC Congress, A Coruña, Spain, 27–28 September 2018.

Published: 17 September 2018

Abstract: The training of deep neural networks usually requires a vast amount of annotated data, which is expensive to obtain in clinical environments. In this work, we propose the use of complementary medical image modalities as an alternative to reduce the required annotated data. The self-supervised training of a reconstruction task between paired multimodal images can be used to learn about the image contents without using any label. Experiments performed with the multimodal setting formed by retinography and fluorescein angiography demonstrate that the proposed task produces the recognition of relevant retinal structures.

Keywords: self-supervised; multimodal; retinography; angiography

1. Introduction

In clinical practice routine, patients are typically subjected to multiple imaging tests, producing complementary visualizations of the same body parts or organs. This leaves available large sets of paired multimodal images. This paired data can be used to train a neural network to predict one modality from other. If the transformation between modalities is complex enough, the network will have to learn about the objects represented in the images to solve the task. This domain-specific knowledge can be used to complement the training of additional tasks in the same application domain, reducing the amount of labeled data required.

We applied the described paradigm to the multimodal image pair formed by retinography and fluorescein angiography. These image modalities are complementary representations of the eye fundus. The angiography has additional information about the vascular structures due to the use of an injected contrast. This also makes this modality invasive and less employed. We train a neural network to predict the angiography from a retinography of the same patient and demonstrate that the network learns about relevant structures of the eye with this self-supervised training [1].

2. Methodology

The multimodal reconstruction is trained with a set of retinography-angiography pairs obtained from the public Isfahan MISP database. This dataset includes 59 image pairs from healthy individuals and from patients diagnosed with diabetic retinopathy.

The multimodal image pairs are aligned following the methodology proposed in [2] to produce a pixel-wise correspondence between modalities. After the image alignment, a reconstruction loss can be directly computed between the network output and the target image. This allows the self-supervised training of the multimodal reconstruction, which will generate a pseudo-angiography representation for any retinography used as input to the network. Three difference functions are considered to obtain the reconstruction loss: L1-norm, L2-norm and SSIM.

The multimodal reconstruction is based on a U-Net network architecture. The network training is performed with the Adam algorithm with an initial learning rate of a = 0.0001, which is reduced by a factor of 0.1 when the validation loss plateaus. Spatial data augmentation is used to reduce the overfitting.

3. Results and Conclusions

Examples of generated pseudo-angiographies are depicted in Figure 1. It is observed that the best results are obtained training with SSIM, in which case the network has learned to adequately transform relevant retinal structures. An additional experiment is performed to specifically measure the ability to recognize the retinal vasculature. A global thresholding is applied to produce a rough vessel segmentation from both the pseudo-angiography and the original retinography. This experiment is performed in the public DRIVE dataset, which comprises 40 retinographies and their ground truth vessel segmentation. The results are evaluated with the Receiver Operator Characteristic (ROC) curves. The measured Area Under Curve (AUC) values are 0.5811 for the retinographies and 0.8183 for the pseudo-angiographies generated after training with SSIM. This improvement demonstrates that the multimodal reconstruction provides additional information about the retinal vasculature.

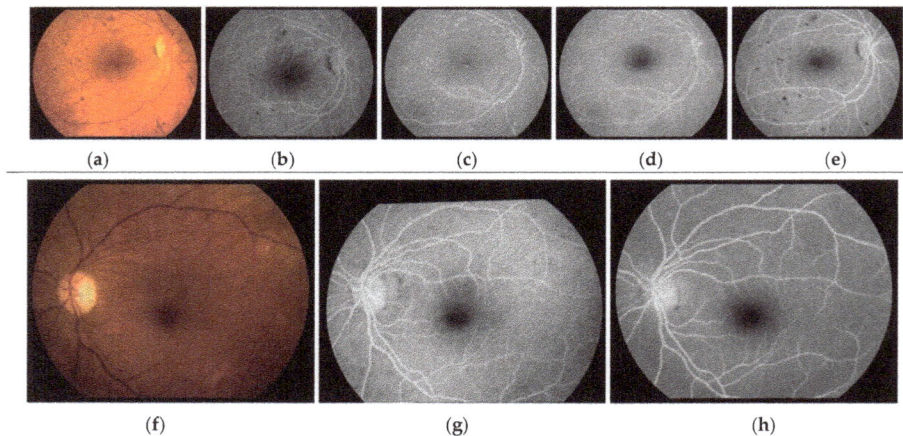

Figure 1. Examples of generated pseudo-angiographies: (**a**,**f**) original retinographies; (**b**) original angiography for (**a**); (**c**–**e**) pseudo-angiographies generated from (**a**) after training with L1 (**c**), L2 (**d**) and SSIM (**e**); (**g**) original angiography for (**f**); (**h**) pseudo-angiography from (**f**) after training with SSIM.

The results indicate that the proposed task can be used to produce a pseudo-angiography representation and learn about the retinal structures without requiring any annotated data.

Author Contributions: A.S.H., J.R. and J.N. contributed to the analysis and design of the computer methods and the experimental evaluation methods, whereas A.S.H. also developed the software and performed the experiments. M.O. contributed with domain-specific knowledge, the collection of images, and part of the registration software. All the authors performed the result analysis. A.S.H. was in charge of writing the manuscript, and all the authors participated in its critical revision and final approval.

Acknowledgments: This work is supported by I.S. Carlos III, Government of Spain, and the ERDF of the EU through the DTS15/00153 research project, and by the MINECO, Government of Spain, through the DPI2015-69948-R research project. The authors of this work also receive financial support from the ERDF and ESF of the EU, and the Xunta de Galicia through Centro Singular de Investigación de Galicia, accreditation 2016-2019, ref. ED431G/01 and Grupo de Referencia Competitiva, ED431C 2016-047 research projects, and the predoctoral grant contract ref. ED481A-2017/328.

Proceedings **2018**, *2*, 1195

Conflicts of Interest: The authors declare no conflict of interest. The founding sponsors had no role in the design of the study; in the collection, analyses, or interpretation of data; in the writing of the manuscript, and in the decision to publish the results.

References

1. Hervella, A.S.; Rouco, J.; Novo, J.; Ortega, M. Retinal Image Understanding Emerges from Self-Supervised Multimodal Reconstruction. In Proceedings of the Medical Image Computing and Computer-Assisted Intervention (MICCAI), Granada, Spain, 16–20 September 2018.
2. Hervella, A.S.; Rouco, J.; Novo, J.; Ortega, M. Multimodal Registration of Retinal Images Using Domain-Specific Landmarks and Vessel Enhancement. In Proceedings of the International Conference on Knowledge-Based and Intelligent Information and Engineering Systems (KES), Belgrade, Serbia, 3–5 September 2018.

proceedings

MDPI

Extended Abstract

Software Defined Radio: A Brief Introduction †

Anxo Tato

atlanTTic Research Center, University of Vigo, 36310 Vigo, Pontevedra, Spain; anxotato@gts.uvigo.es;
Tel.: +34-986-812-000

† Presented at the XoveTIC Congress, A Coruña, Spain, 27–28 September 2018.

Published: 19 September 2018

Abstract: In this short article the concept of Software Defined Radio (SDR) is introduced and compared with the traditional radio. Then, a research project of atlanTTic center which used this technology was briefly presented and lastly, we include a reference to some dissemination activities related with SDR to be developed shortly.

Keywords: telecommunications; satellite communications; software defined radio (SDR)

1. Introduction

A radio is a system with technology for transferring information wirelessly by means of electromagnetic radiation [1]. In the past a radio was composed of many discrete circuits and electronic devices and it had a fixed functionality which could not be modified after manufacturing. For example, with a traditional radio one could not turn a commercial FM receiver into a digital radio receiver. However, nowadays with the Software Defined Radio (SDR) one can buy an USB DVB-T2 dongle designed for reception of terrestrial TV in a computer and use it as a GPS receiver, or to decode ADS-B (Automatic Dependent Surveillance-Broadcast) signals and obtain the positions of all the planes in the nearby. This shows how a SDR outperforms a traditional radio in terms of flexibility and reconfigurability.

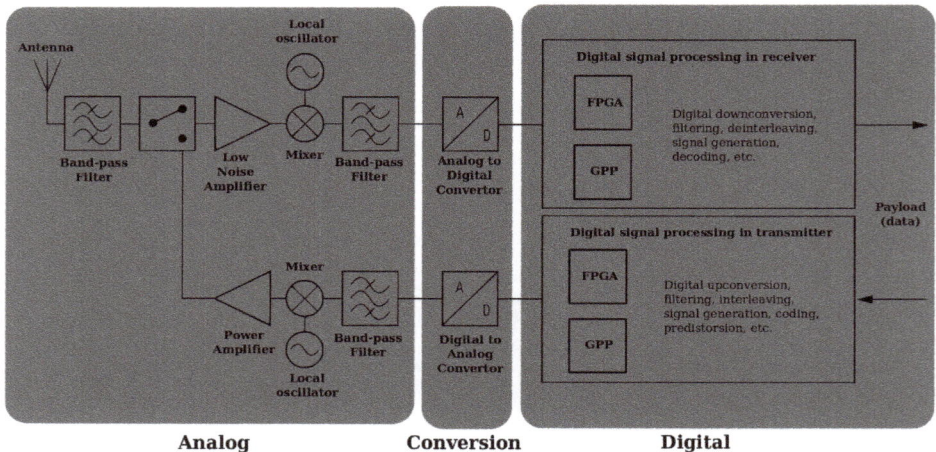

Figure 1. Possible design of a software defined radio. Adapted from [2].

According to the Wireless Innovation Forum a SDR is a radio in which all or part of the functions of the physical layer are defined by software. The increase in the power of the processors and FPGAs, the reduction of its price and consumption and the emergence of various integrated RF transceiver circuits made it possible for this technology to grow in recent years extending from military and research to commercial and amateur systems. Figure 1 shows a simplified block diagram of a SDR. In the middle we have an Analog to Digital Convertor (ADC) which transforms real world analogue signals into digital and discrete signals which a digital circuit can process and a Digital to Analog Convertor (DAC) which transforms digital samples into an analogue waveform to feed the Radio Frequency (RF) stage which prepares the signal to be transmitted by the antenna.

The last component of a SDR is typically a General Purpose Processor (GPP) where all the digital signal processing takes place. Therefore, it is possible to modify many parameters of the physical layer of a system (or even change all the physical layer!) since it is software defined. In the same way as we install different programs or applications in a computer or smartphone and we give it a new functionality, changing the software which is run in the GPP we can convert the radio into a communication system which follows any given standard as Bluetooth, WiFi, FM, DVB-T2, GSM, LTE, etc., or even in a system with an arbitrary user-defined waveform as is sometimes done in research, for example. A SDR, due to its flexibility, is considered an enabling technology for advanced communication systems which require some type of reconfiguration capability as adaptive or cognitive radio.

2. atlanTTic Experience with SDR

In atlanTTic, the research center for Telecommunication Technologies promoted by the University of Vigo (Galicia, Spain), we have been using SDR in some of our research projects. Recently, in a project named Tactica [3] this technology was used to validate algorithms for adaptive communications in a real scenario, see Figure 2. Using two SDR platforms, a two-way link was established communicating a mobile terminal (embarked on a car or a fixed-wing Unmanned Aerial Vehicle, UAV) with a base station through a medium orbit (MEO) satellite.

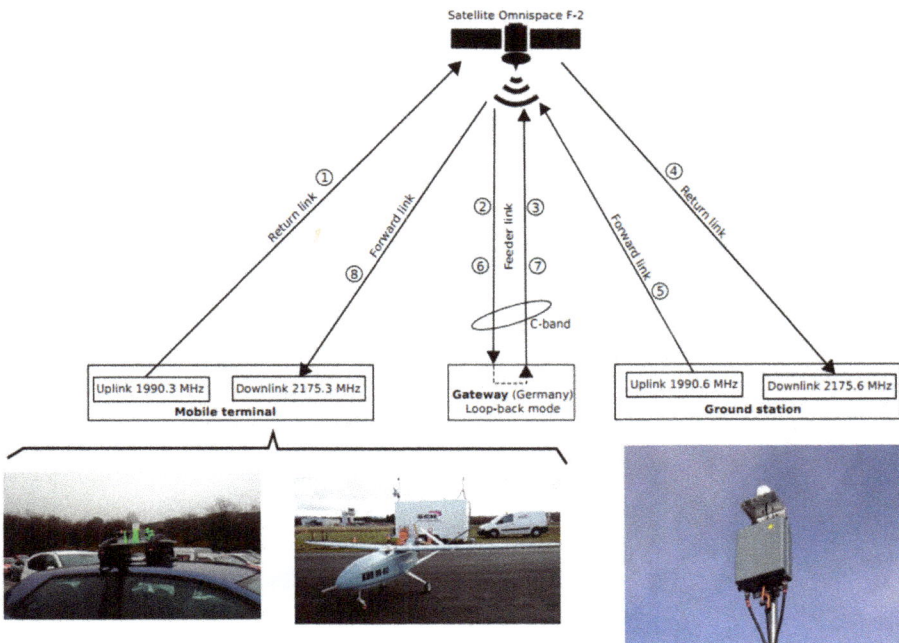

Figure 2. Block diagrams of the satellite communications system created in Tactica project.

115

The SDR platform selected for the project was the USRP Ettus E310 which has a Xilinx Zynq 7020 SoC (System on Chip) and it allowed the system to operate in real time to send up to approximately 200 kbps when the 16-QAM constellation was selected. The physical layer of the communications system was based on the standard ETSI TS 102 704, a specification of the satellite component of 3G cellular networks which is used by the BGAN service of Inmarsat. In this standard the transmitter is able to adapt its parameters to follow the varying channel conditions, being this able to modify the bandwidth, the transmitted power, the modulation and also the coding rate of the channel encoder. The SDR technology allowed us to develop more easily a prototype of an adaptive satellite communications system and test different algorithms to select the proper configuration of the physical layer automatically without any human intervention.

3. Dissemination Activities Related with SDR

In the developed countries companies are facing difficulties in recruiting ICT skilled workers and this problem is even worst when they look for engineers specialized in communications and signal processing. Moreover, in our universities students also do not find the latter specialities so appealing. Knowing this, a group of engineers of the atlanTTic and CINAE research centers has started a projected for realizing dissemination activities of the SDR technology, see Figure 3. This project has founding from Iniciativa Xove, a program of the Youth Department of the Galicia Government.

The project [4] would allow to introduce the SDR technology to young students of engineering by means of a conference where a senior researcher will explain it from scratch with examples of his recent projects and developments. It is also foreseen the realization of a four-day workshop where the participants can do some hands-on projects with the SDR boards LimeSDR Mini and GNU Radio framework. The ultimate goal is to arouse their curiosity about wireless communication and put in practice with real hardware what they learn in the university courses. On the long term, we would like to encourage the creation of a community of people who works and experiments with SDR in the same way as there are also developers groups around other software technologies and programming languages.

Figure 3. Poster of the dissemination activities related with SDR.

Acknowledgments: This work was funded by the Xunta de Galicia (Secretaria Xeral de Universidades) under a predoctoral scholarship (co-funded by the European Social Fund).

Conflicts of Interest: The author declares no conflict of interest. The founding sponsors had no role in the design of the study; in the collection, analyses, or interpretation of data; in the writing of the manuscript, and in the decision to publish the results.

References

1. Collins, T.F.; Getz, R.; Pu, D.; Wyglinski, A.M. *Software-Defined Radio for Engineers*; Artech House: Norwood, MA, USA, 2018.
2. Software Defined Radios—Overview and Hardware. Available online: https://cdn.rohde-schwarz.com/pws/dl_downloads/dl_common_library/dl_news_from_rs/182/n182_radiocomunit.pdf (accessed on 7 September 2018).
3. Tato, A.; Mosquera, C.; Gomez, I. Link adaptation in mobile satellite links: Field trials results. In Proceedings of the 8th Advanced Satellite Multimedia Systems Conference and the 14th Signal Processing for Space Communications Workshop (ASMS/SPSC), Palma de Mallorca, Spain, 5–7 September 2016.
4. SDR Galicia. Available online: https://sdrgal.wordpress.com (accessed on 7 September 2018).

proceedings

Extended Abstract
Texture Mapping on NURBS Surface [†]

Sergio Vázquez [1,2,*] and Margarita Amor [2]

[1] CITIC—Research Center of Information and Communication Technologies, Universidade da Coruña, Elviña, 15071 A Coruña, Spain
[2] Departamento de Ingeniería de Computadores, Universidade da Coruña, Elviña, 15071 A Coruña, Spain; margarita.amor@udc.es
* Correspondence: sergio.vazquez@udc.es; Tel.: +34-981-167-000
† Presented at the XoveTIC Congress, A Coruña, Spain, 27–28 September 2018.

Published: 17 September 2018

Abstract: Texture mapping allows high resolution details over 3D surfaces. Nevertheless, texture mapping has a number of unresolved problems such as distortion, boundary between textures or filtering. On the other hand, NURBS surfaces are usually decomposed into a set of Bézier surfaces, since NURBS surface can not be directly rendered by GPU. In this work, we propose a texture mapping directly on the NURBS surfaces using the RPNS (Rendering Pipeline for NURBS Surface) method, which allows the rendering of NURBS surface directly on the GPU. Our proposal facilitates the implementation while minimizing the cost of storage, mitigating distortions and stitching between textures.

Keywords: NURBS; texture; shader; GPU

1. Introduction

NURBS (Non-Uniform Rational B-Splines) [1] surfaces are one of the standards for data representation, design and exchange in CAD/CAM/CAE applications. NURBS surfaces with textures allows a more realistic representation of the surfaces, improving the final scene in areas such as modeling, virtual reality or animation [2]. Texture mapping presents a set of problems that usually require user intervention [3]. The application of texture NURBS surfaces implies a high cost of storage due to the utilization of techniques such as texture atlas generation [4]. In this work we propose a texture mapping directly on the NURBS surfaces using RPNS (Rendering Pipeline for NURBS Surface) [5]. RPNS is a solution for the direct evaluation of NURBs surface on the GPU without any previous decomposition to Bézier surfaces.

2. Texture Rendering

RPNS adds a new primitive to the input flow of the geometry stage, KSQuad. In addition, an intermediate stage, the sampler, is added between the geometry and the rasterization stages, as shown in Figure 1. In this stage, KSQuad primitives are sampled adaptively according to the point of view, the geometric characteristics of the surface, and the contour edges between surfaces. This sampling results in a set of sampling points or dices called KSDice that allow the surface to be rendered without cracks or holes. Each KSDice consists of a sampling point and additional information such as the parametric size of the matrix and the grade of the corresponding surface, and does not store any explicit connectivity information. KSDice is a primitive that can ultimately be projected to a single pixel or a set of pixels.

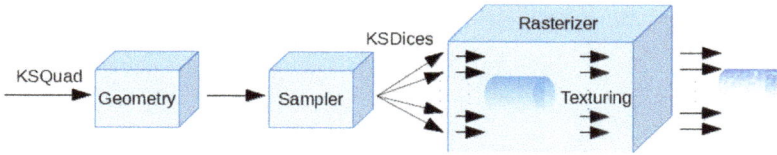

Figure 1. Rendering texture mapping pipeline for NURBS surfaces.

In the implementation of our proposal, we make use of Texture2DArray, a type of texture coordinated with DirectX and HLSL5 (High Level Shader Language), that allows the definition of colors directly on the KSDices.

3. Results and Conclusions

In this section we present the results obtained on different models (Figure 2) with our proposal. The platform on which the different tests were performed consists of an Intel i7-4790 3.6 GHz with 32 GB of RAM and a NVidia GTX 1080ti with 11GB GDDR5X. As for the software, the tests were conducted on Windows 10 using Visual Studio Community 2017 with DirectX 11 and Microsoft HLSL.

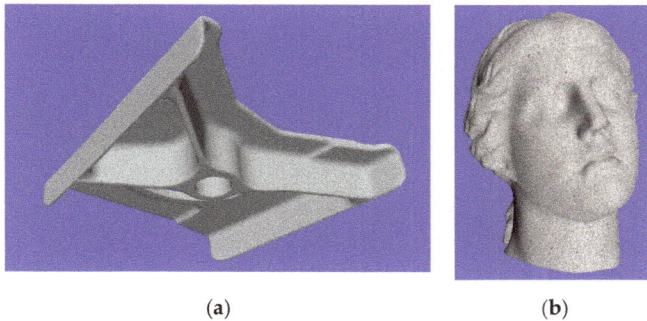

(**a**)   (**b**)

Figure 2. NURBS Models: (**a**) Hinge; (**b**) Head.

Our proposal shows how NURBS surface can be used with textures without previous transformations.

Author Contributions: All authors contributed equally.

Acknowledgments: This research was funded by [the Ministry of Economy and Competitiveness of Spain and FEDER funds (80%)] grant number [TIN2016-75845-P], by [the Government of Galicia (Xunta de Galicia) co-founded by ERDF funds under the Consolidation Programme of Competitive Reference Groups] grant number [ED431C 2017/04], by [the Consolidation Programme of Competitive Research Units] grant number [R2014/049 and R2016/037] and by [the Xunta de Galicia (Centro Singular de Investigación de Galicia accreditation 2016–2019) and the European Union (European Regional Development Fund, ERDF)] grant number [ED431G/01].

Conflicts of Interest: The authors declare no conflict of interest. The founding sponsors had no role in the design of the study; in the collection, analyses, or interpretation of data; in the writing of the manuscript, and in the decision to publish the results.

References

1. Les Piegl, W.T. *The NURBS Book*, 2nd ed.; Springer: Berlin, Germany, 1997.
2. Pascu, N.E.; Dobrescu, T.; Opran, C.; Enciu, G. Realistic Scenes in CAD Application. *Procedia Eng.* **2014**, *69*, 304–309.
3. Yuksel, C.; Keyser, J.; House, D.H. Mesh colors. *ACM Trans. Gr.* **2010**, *29*, 15.

4. Guthe, M.; Klein, R. Automatic Texture Atlas Generation from Trimmed NURBS Models. *Comput. Gr. Forum* **2003**, *22*, 453–462.

5. Concheiro, R.; Amor, M.; Padrón, E.J.; Doggett, M. Interactive rendering of NURBS surfaces. *Comput. Aided Des.* **2014**, *56*, 34–44.

proceedings

MDPI

Extended Abstracts

Technologies for Participatory Medicine and Health Promotion in the Elderly Population [†]

Naveira-Carro Eloy *, Concheiro-Moscoso Patricia * and Miranda-Duro MC *

Research group Artificial Neural Networks and Adaptative, Systems-Center of Medical Informatics and Radiological Diagnosis, (RNASA-IMEDIR), Research Center on Information and Communication, Technologies (CITIC), Faculty of Health Sciences, Universidade da, Coruña, 15071 A Coruña, Spain

* Correspondence: eloy.naveira@udc.es (N.-C.E.); patricia.concheiro@udc.es (C.-M.P.); carmen.miranda@udc.es (M.-D.M.)

† Presented at the XoveTIC Congress, A Coruña, Spain, 27–28 September 2018.

Published: 28 September 2018

Keywords: participatory medicine; technological development; elderly

1. Introduction

The progressive aging of the population is a socio-demographic phenomenon experienced by most countries in the world in recent decades, especially in Japan and in many European Union countries. During this process, so-called "geriatric syndromes" frequently occur. The focus of this study is the quality of life of the elderly in relation to these three factors: risk of falls, urinary incontinence, and sleep disorders.

This project is based on the use of wearables devices, which used for the measurement of different biomedical parameters that serve to monitor and analyze aspects such as sleep, physical activity, among others, which favors the monitoring of people during an investigation. In addition, this project is developing a web application where people register daily aspects of their daily occupations. Both the use of wearables and the web registry favor participatory medicine, so that people are active agents in the management of their own health.

In the field of health, more and more technology companies are betting on the development of sensor devices and applications for patient monitoring, which allows a detailed monitoring of the health of users, with its consequent benefits. By using these devices, we can quantify movements and body parameters.

2. Objectives

The main purpose is to determine the impact of a multifactorial intervention program implemented with institutionalized elderly people. The program is focused on the treatment of the aforementioned factors.

3. Material and Methods

The study will be carried out with elderly people living in three residences for the elderly in A Coruña Province (Galicia, Spain).

It is a prospective and longitudinal study, with a temporary series design of a "quasi-experimental" type that evaluates the effect of an intervention in one given population by doing assessments pre- and post-intervention, but there is no comparison with a control group.

The intervention will be based on a multifactorial program, including the following phases: the use of wearable devices (wearable fitness trackers to register physical activity and sleep), the use of an App on a Tablet to record the participants' occupations and activities, counseling about performance in activities of daily living, the implementation of a physical activity program, and the treatment of the pelvic floor (according to each research line). The Quality of Life (QoL) will be assessed before and after the intervention, with the use of the questionnaire EuroQol-5D-5L. Data analysis will be applied with all registered variables through a quantitative perspective.

4. Results and Conclusions

Due to previous experiences with similar projects to the one presented here, this project can contribute to the reduction of the signs and symptoms of the syndromes: urinary incontinence, risk of falling and sleep disturbances. In addition, with the advice offered to the participants, they are training themselves in case of having a problem, reduce the consequences. We will have to wait for the complete analysis of the results in order to draw a conclusion in accordance with the data obtained, but as observed, there is a great acceptance of the program by the participants.

The program will continue to be implemented with new participants to ensure its relevance and validity in different contexts and people profiles.

Authors Contributions: This Project has been developed by a multidisciplinary team. Each of the professionals included has been responsible for the execution of each of the parts of this investigation. So, E.N., who is a computer scientist, has been the person in charge of the development of the technological platform of said project. While, P.C. and M.d.C.M., both occupational therapists, have focused on the design and implementation of the technological platform and the intervention carried out in the study together with other professionals. For the presentation, both have contributed to the design, preparation and writing of the same.

Acknowledgments: GERIA-TIC Project, Project co-funded by the Galician Innovation Agency (GAIN) through the Connect PEME Program (3rd edition) (IN852A 2016/10) and EU FEDER funds, Collaborative Genomic Data Integration Project (CICLOGEN). Data mining and molecular docking techniques for integrative data analysis in colon cancer. "Funded by the Ministry of Economy, Industry and Competitiveness. Galician Network of Research in Colorectal Cancer (REGICC) ED431D 2017/23, Galician Network of Medicines (REGID) ED431D 2017/16 funded by the Department of Culture Education and University Planning aids for the consolidation and structuring of competitive research units of the University System of Galicia of the Xunta de Galicia and Singular Centers (ED431G/01) endowed with FEDER funds of the EU.

Conflict of Interests: The authors declare no conflict of interest.

Extended Abstract

Fully Automatic Teeth Segmentation in Adult OPG Images [†]

Nicolás Vila Blanco [1],*, Inmaculada Tomás Carmona [2] and María José Carreira Nouche [1]

[1] Centro de Investigación en Tecnoloxías da Información (CITIUS), Universidade de Santiago de Compostela, 15782 Santiago de Compostela, Spain; inmaculada.tomas@usc.es (I.T.C.), mariajose.carreira@usc.es (M.J.C.N.)

[2] Oral Sciences Research Group, Universidade de Santiago de Compostela, Health Research Institute Foundation of Santiago (FIDIS), 15872 Santiago de Compostela, Spain; nicolas.vila@usc.es

* Correspondence: nicolas.vila@usc.es; Tel.: +34-8818-16445

† Presented at the XoveTIC Congress, A Coruña, Spain, 27–28 September 2018.

Published: 17 September 2018

Abstract: In this work, the problem of segmenting teeth in panoramic dental images is addressed. The Random Forest Regression Voting Constrained Local Models (RFRV-CLM) are used to perform the segmentation in two steps. Firstly, a set of mandible and teeth keypoints are located, and then that points are used to initialise each individual tooth model. A method to detect missing teeth based on the quality of fit is presented. The system is evaluated using 346 manually annotated images containing adult-stage teeth. Encouraging results on detecting missing teeth are achieved. The system is able to locate the outline of the teeth to a median point-to-curve error of 0.2 mm.

Keywords: teeth segmentation; panoramic dental images; random forest regression-voting; machine learning

1. Introduction

Since they discovery, dental X-ray images have been widely used in a variety of clinical fields, such as abnormality detection, treatment and surgery planning, prostheses design, assessment of children's dental development, human identification and many more. Extraoral panoramic images in particular show a full coverage of the teeth as well as other surrounding bones, such as the mandible or the vertebrae. However, the quality of these images is quite challenging to automatic processing algorithms, mainly because the acquisition process is highly dependent on the patient positioning and patient movements.

2. Methods

Our main contribution is the development of a fully automatic procedure to detect and outline mandibular adult-stage teeth in panoramic dental images, and a simple method to detect missing teeth. To do that, Random Forest Regression Voting Constrained Local Models (RFRV-CLM) are used. This method combines a global linear shape model with local appearance models to locate each shape point. Full details of the method are explained in [1].

One possible approach is to build an individual model for each specific tooth. Due to the symmetry of the mouth, the teeth models of one mandible side can be used to outline the teeth on the other side. However, there are two main limitations. First of all, the search space is too big when compared with the target teeth shapes, so the teeth models needs a reasonably good initialisation. Furthermore, the teeth are very close to each other and the shapes are very similar within each tooth type (incisors, molars, etc.), so the tooth search can easily converge to a neighbouring tooth.

To overcome this problem, a two-step segmentation procedure is proposed. In the first step, a RFRV-CLM model is trained to detect a set of mandible and keypoints. This allows to capture the pose variation of teeth in relation with other teeth and with the mandible. Furthermore, the model initialization is easier due to the target shape occupies the great part of the image. In the second step, the initial shapes for each tooth model are calculated from the previous detected keypoints, and refined with the individual teeth models. Besides, a simple method based on the thresholding of the quality-of-fit per tooth is applied after the teeth shape search in order to detect missing teeth.

The full procedure was evaluated in a set of 346 panoramic images (261 images for training and 85 for testing). In each image, the shapes of seven left-mandibular teeth were manually annotated. The individual tooth models and the keypoint model were built with the RFRV-CLM algorithm. The predicted shapes of left-mandibular teeth were compared to ground truth and the performance was assessed in two ways.

Firstly, the missing teeth detection was evaluated as a classification problem with two target classes: missing (negative class) or present (positive class). The accuracy of the system was over 95%, where the precision, sensitivity and specificity were 99%, 96% and 84%, respectively.

Secondly, the accuracy of the teeth shape outlining was assessed with the point-to-curve error, which represents the shortest distance of each predicted shape point to the curve through the ground truth points. This measurement was obtained in correctly located teeth, i.e., the teeth whose predicted shapes overlaps with the ground truth shape more than 50%. The results show a median error of less than 0.23 mm for all types of teeth and the 99% ile is 1.31 mm in the worst case, which demonstrate the robustness of this procedure.

Author Contributions: N.V.B. designed and developed the experiments, analysed the results and wrote the paper. I.T.C. provided the OPG database and validated the results from a clinical point of view. M.J.C.N. helped to design the experiments and analyse the results, and validated the results from a technical point of view.

Acknowledgments: This work has received financial support from the Consellería de Cultura, Educación e Ordenación Universitaria (accreditation 2016–2019, ED431G/08; ED431B 2017/029) and the European Regional Development Fund (ERDF).

Reference

1.   Lindner, C.; Bromiley, P.A.; Ionita, M.C.; Cootes, T.F. Robust and accurate shape model matching using random forest regression-voting. *IEEE Trans. Pattern Anal.* **2015**, *37*, 1862–1874.

MDPI

St. Alban-Anlage 66

4052 Basel

Switzerland

Tel. +41 61 683 77 34

Fax +41 61 302 89 18

www.mdpi.com

Proceedings Editorial Office

E-mail: proceedings@mdpi.com

www.mdpi.com/journal/proceedings

www.ingramcontent.com/pod-product-compliance
Lightning Source LLC
Chambersburg PA
CBHW051911210326
41597CB00033B/6103